032053

Springer Series in Perception Engineering

Series Editor: Ramesh Jain

Springer Series
in Perception Engineering

P.J. Besl: *Surfaces in Range Image Understanding*

J.L.C. Sanz (ed.): *Advances in Machine Vision*

R.C. Vogt: *Automatic Generation of Morphological Set Recognition Algorithms*

Robert C. Vogt

Automatic Generation of Morphological Set Recognition Algorithms

With 96 Illustrations

Springer-Verlag
New York Berlin Heidelberg
London Paris Tokyo Hong Kong

Robert C. Vogt
ERIM
Image Processing Systems Division
P.O. Box 8618
Ann Arbor, MI 48107
USA

Series Editor
Ramesh Jain
Electrical Engineering
and Computer Science Department
The University of Michigan
1101 Beal Avenue
Ann Arbor, MI 48109
USA

Library of Congress Cataloging-in-Publication Data
Vogt, Robert C. (Robert Carl)
Automatic generation of morphological set recognition algorithms / Robert C. Vogt.
p. cm.—(Springer series in perception engineering)
Bibliography: p.
ISBN 0-387-97049-5 (alk. paper)
1. Image processing—Digital techniques. I. Title. II. Series.
TA1632.V64 1989
621.36'7—dc20 89-11306

Printed on acid-free paper.

Camera-ready copy provided by the author.
Printed and bound by Edwards Brothers, Inc., Ann Arbor, Michigan.
Printed in the United States of America.

9 8 7 6 5 4 3 2 1

ISBN 0-387-97049-5 Springer-Verlag New York Berlin Heidelberg
ISBN 3-540-97049-5 Springer-Verlag Berlin Heidelberg New York

For Marie Jo and Robin

Series Preface

Since the early days of computers, machine learning and automatic programming have attracted researchers in computer science and related fields, particularly pattern recognition and automatic control theory. Most of the learning concepts in machine perception have been inspired by pattern recognition approaches that rely on statistical techniques. These statistical techniques have applicability in limited recognition tasks. Automatic programming in perception systems has generally been limited to interfaces that allow easy specification of the task using natural language. Clearly, machine learning and automatic programming can make perception systems powerful and easy to use.

Vogt's book addresses both these tasks in the context of machine vision. He uses morphological operations to implement his approach which was developed for solving the figure-ground problem in images. His system selects the correct sequence of operators to accept or reject pixels for finding objects in an image.

The sequence of operators is selected after a user specifies what the *correct* objects are. On the surface it may appear that the problem solved by the system is not very interesting, however, the contribution of Vogt's work should not be judged by the images that the system can segment. Its real contribution is in demonstrating, possibly for the first time, that automatic programming is possible in computer vision systems. The selection of morphological operators demonstrates that to implement an automatic programming-based approach, operators whose behavior is clearly defined in the image space are required. This book demonstrates that many early operations in machine vision systems may benefit from similar approaches that will allow *training* systems to perform in an application domain. This training can be done using interactive graphics tools by a user unfamiliar with machine vision technology.

This book addresses an aspect of machine perception that will become increasingly important as the field matures. It will be useful to practitioners developing application systems and will inspire researchers to explore more powerful approaches to learning and automatic programming in machine perception systems.

Ramesh Jain
February, 1989

Preface

One of the biggest problems in the field of machine vision is that algorithm development is a slow, time consuming task for which the level of available human expertise does not meet the demand. In the early 80's it was thought that simply being able to build computers that could operate on 512x512 images at video rates would be enough to pay for the commercialization of computer vision. Over the past several years, with the difficulties experienced by some of the start-up companies in this field, it has become clear that that, in itself, was not enough. The ability to *execute* a known algorithm quickly does not by itself solve the more difficult problem of how to *find* an algorithm that works sufficiently well within a reasonable time. The finding of algorithms (i. e., the development of reliable image processing software) requires searching through a rather complicated space of operators and parameters. The time required for this activity has turned out to be a major stumbling block hampering progress in the field.

At the same time, the process of image algorithm development is more time consuming and less motivating than it perhaps could be, because it typically requires repeating the same algorithm sequences over many different images. It is also mentally taxing because the developer has to visually evaluate and mentally integrate how well each algorithm performs over the entire image set, and from this propose changes in parameters or operator sequences which will improve the overall performance.

In my own early experience with image algorithm development, I felt that much of what I was doing was quite mechanical, and could probably be done better and faster by a machine, if only an appropriate system architecture were available for representing algorithms, running tests, evaluating results, and making decisions.

The work being reported here describes a program, called "**REM**", which I have designed and implemented in order to study the process of image algorithm development, originally as part of my doctoral thesis in computer science at the University of Michigan. To be more specific, the work is concerned with automating the programming of image processing computers for set recognition tasks. There are several reasons why research into this area is very important. First of all, the process of formalizing the use of the image operators is quite valuable. This forces one to explicitly define the conditions for which a given operator will succeed and for which it should be considered. Much of this knowledge is not written down anywhere; it is simply acquired with experience by those who work in the field.

Second, research into this area can help us to improve software development productivity for imaging problems. Automating the process of algorithm development requires writing routines which can test whole sets of images at a time, or make measurements on images related to the effects of each operator, or evaluate the result of applying an operator on an image, or summarize results over a series of images toward the goal of solving a problem (among other tasks). The same kinds of routines can be developed into better, more powerful tools for use by image algorithm developers. This offers the potential for faster development cycles for image algorithms, improved algorithm accuracy and robustness, and better utilization of both humans and image processing machines. By off-loading some of the more repetitious and boring tasks of algorithm development to machines, humans could spend more time thinking about new approaches, and less time executing and evaluating them. This should increase both their motivation and productivity.

The third reason for doing research into automating image algorithm development, beyond improving our understanding of the operators and the development of more powerful tools, is to produce systems which can manage the actual search for an algorithm. This is also a difficult, error-prone task for humans, owing to the the size of the search and the need to record and compare observations on many algorithms over large numbers of images. The idea proposed here is to incorporate expert knowledge about the use of the operators, about how to evaluate and compare algorithms, and about algorithm development strategies, into a computer-accessible knowledge base. This knowledge base, along with the tools described above, plus additional machinery for representing algorithm states, search trees, and evaluation data about algorithms, would make it possible for a machine to conduct the search entirely automatically. Admittedly, this is a longer-term goal, which will require step-by-step development, starting with very simple, idealistic problems before moving on to more complex and realistic ones. As this book shows, however, we can make progress here too.

In general, the problems encountered in developing image algorithms are quite hard. One typically has samples of tens or hundreds of images with which to work, each of which contains hundreds of thousands of pixels. The objects to be recognized may look very different from image to image, even when there is no third dimension to consider. There are many operators to choose from, each of which may have tens or hundreds of different possible parameter settings. Since most of these operators do not commute with each other, order is important, and thus the number of possible operator sequences explodes rather rapidly, even when the sequences are short. Moreover, the length of time it takes to run algorithm sequences over a large number of 512x512 sample images (even on a high speed image processor), makes it difficult to even consider sequences that are very long. Indeed, if it were not for the fact that our own visual systems tell us the images are sufficiently well-behaved to recognize the objects of interest, the chances of coming up with algorithms that worked well on entire image sets by any un-intelligent method would be so slim, that it would probably preclude doing any work in this area at all! This is precisely why anything we can do to automate the process by off-loading repetitive testing and evaluation tasks to machines, should be welcomed by practitioners in the field. Research into these questions deserves serious consideration and support, particularly since most of the tasks being discussed are ones that we already know machines can perform well.

The REM system described here is written in Vax Common Lisp and runs on a Vax 8650 under the VMS operating system (see Figure 1.2).[1] Its image processing tasks are carried out by a Cytocomputer-HSS, the fourth generation of Cytocomputer image processors designed and built at the Environmental Research Institute of Michigan (ERIM) [Lougheed 1985,1987]. The Cyto-HSS is linked to the Vax via an Ethernet connection. We are able to drive the Cyto hardware from Lisp by means of a software interface we have established between it and C4PL, the interactive image software development environment (also developed at ERIM), which is normally used to program the Cytocomputer [Lougheed & Overmars 1987].[2]

While the REM program represents only an initial exploration into this area, and is fairly limited in the breadth and complexity of problems that it can solve, it does contain a well-developed model of the algorithm development process which can be easily augmented with additional expertise and capabilities. It can already solve simple problems from a repertoire of several thousand operator/parameter combinations. It suggests that we can make significant progress in speeding up algorithm development by studying and formalizing how that process is normally carried out by humans.

R. C. Vogt

Ann Arbor
September 1988

---

[1]'Vax' and 'VMS' are trademarks of the Digital Equipment Corporation.

[2]'Cytocomputer', 'Cyto-HSS', and 'C4PL' are trademarks of the Environmental Research Institute of Michigan.

Acknowledgements

The help, efforts and support of many people were instrumental to the completion of this work. Dr. Stanley Sternberg first introduced me to image processing and mathematical morphlogy, which was really the beginning of this work as well as my career. He also introduced me to Dr. Jean Serra, Director of the Centre de Morphologie Mathematique at the Ecole des Mines in Fontainebleau, France—without a doubt the world's leading expert in this area. Dr. Serra has done probably more than anyone to help me accomplish the technical goals of this research. I am grateful to him for his help in making it possible for me to work in France in this subject area, and to interact with his group at the Ecole des Mines. I also wish to thank him for the many fruitful discussions we had during the 2 1/2 years that I was there.

Thanks are also due to Thomson-CGR, France, where I worked for 2 years, and in particular to my supervisor, Dr. Claude Benchimol, who made it possible for me to work on interesting medical imaging problems, while at the same time pursuing my Ph.D. research. While there, I was able to develop Lisp programs and image processing algorithms that later became the basis of the thesis implementation effort. The experience I gained at CGR in developing morphology algorithms has also proved to be quite valuable for my later work and research.

Likewise, I wish to express my thanks to the Environmental Research Institute of Michigan, in Ann Arbor, and particularly to Mr. Ron Swonger, Dr. Robert Sampson, and Dr. Robert Lyjak, for their support of my research and its implementation, here in the United States. Thanks are also due to the personnel of the Image Processing Technology Laboratory, whose continuing improvements to the C4PL language and to the Cyto-HSS hardware made it possible to do many things in the implementation which had not been envisioned at the beginning. Dr. Charles Jacobus deserves thanks for his support of the development of a Lisp/Cytocomputer interface, as does Cathryn Belisle, who implemented it. This was used heavily in the thesis implementation effort, and made it possible to solve problems in minutes, using the Cytocomputer, rather than in the hours it would have taken using Lisp alone on the Vax.

Some credit is going to have to go to Don Knuth, the author of TeX, and Leslie Lamport, author of LaTeX, as well as to the authors of the University of Michigan Rackham macros, for without all of their efforts it never would have been possible to prepare my thesis and this book in such a short time, particularly with all of the equations and pictures that it contains. The TeXperts at ERIM, in particular Arthur Na and Steve Smith, deserve thanks for answering all of my LaTeX questions and for helping to solve my document preparation problems quickly. I am grateful to Peter Tchoryk for providing a decent editor for my home computer, which made it possible for me to complete the thesis text by the required deadline. Credit is also due to Brian Forester, Jan Ferguson and the ERIM Art Department, who helped in the preparation of the figures, and to the ERIM Photo Lab, which prepared most of the color and black and white prints, and provided others from their archives.

I wish in addition to thank the five members of my thesis committee for their time, efforts, and good suggestions. In particular, thanks are due to Dr. Paul Scott who provided support for me as a graduate student in the area of machine learning, and who served as my committee chair during the first part of my thesis work, as well as to Dr. Ramesh Jain, the later chair, who provided the necessary *push* for me to get it done.

Finally I wish to express my gratitude to my wife, Marie Jo, who gave up her career and came to the United States so that I could finish my doctorate, and who has provided continuous moral support and advice, throughout the many long years of this effort. She always believed in me, even when I did not. To her and to my son, Robin, I owe many many missed evenings and weekends.

Contents

List of Figures

Notation

Objects:

$\mathcal{X}$	Sample Space of Images
$\mathcal{A}$	Sample Space of Accept Masks
$\mathcal{R}$	Sample Space of Reject Masks
X_i	Example Image i
A_i	Accept Mask i
R_i	Reject Mask i
X, Y	Images/Sets
p, q	Points in Images/Set Elements
S, T	Structuring Elements
S'	S Transpose
$S^{\wedge}$	S Reflect
$\mathcal{O}$	Origin
s, t	Points in Structuring Elements
X_t	X Translated by t
X^c	X Complement
$\mathcal{U}$	Universal Image/Set
$\emptyset$	Empty Image/Set

Combination, Mask and Threshold Operations:

$\cup, \bigcup$	Union
$\cap, \bigcap$	Intersection
$\setminus$	Set Subtraction
$\|\|$	Symmetric Difference
$\|$	Concatentate
min,MIN	Minimum
max,MAX	Maximum
$\boxplus$	Clipped Addition
$\boxminus$	Clipped Subtraction
$\perp^x$	Infmask-X
$\perp^y$	Infmask-Y
$\top_x$	Supmask-X
$\top_y$	Supmask-Y
$\|_{L1}^{L2}$	Band Threshold
$\|_L^+$	Upper Threshold
$\|_-^L$	Lower Threshold

Morphological Operations:

Ψ	Arbitrary Base Operation
Ψ_B	Arbitrary Band Operation
$\oplus$	Minkowski Addition
$\ominus$	Minkowski Subtraction
$\circledvee$	Dilation
$\circledwedge$	Erosion
$\overset{r}{\oplus}$	Recursive Dilation
$\overset{r}{\ominus}$	Recursive Erosion
$\circledcirc$	Opening
$\odot$ (filled)	Closing
$\overset{\oplus}{\rightarrow}$	Conditional Dilation
$\overset{\ominus}{\rightarrow}$	Conditional Erosion
$\triangleright$	Reduce
$\triangleleft$	Expand
$\circledast$	Hit or Miss Transform
$\bigcirc$	Thinning
$\odot$	Thickening

Logic, Relationships:

$\forall$	For all
$\exists$	There exists
$\Longrightarrow$	Implies
$\Longleftrightarrow$	If and Only If
$\wedge$	And
$\vee$	Or
$\supseteq$	Contains
$\subseteq$	Is Contained By
$\in$	Belongs To
$\Uparrow$	Hits (Non-Nil Intersection)
$\not\Uparrow$	Misses (Nil Intersection)
Sup	Upper Bound
Inf	Lower Bound

1

Introduction

The project described here was motivated by months of my sitting in front of an image processing system, trying different parameters and operations on a series of similar images, in the hope of developing one algorithm to work for all cases. (The problem was to try to locate the blood vessels in digital coronary angiograms). I remarked that I was spending an enormous amount of time computing partial results for each example image, testing the applicability of operations to subproblems, and looking for ideal parameter ranges over all of the examples considered together. It seemed to me that much of this work could be automated, simply by developing high level routines to do testing and parameter searches on a whole series of examples at once, rather than having me test each one 'by hand'. Going further, it seemed that it should be possible to formalize the applicability of operations to a particular problem, and beyond that to formalize the entire search process for an algorithm. That is what this book is about, and it essentially arose out of my own experience and frustrations at performing these same steps manually for a period of two years while at Thomson-CGR in France, and subsequently at Machine Vision International (MVI) and at the Environmental Research Institute of Michigan (ERIM) in the United States.

Something else that I remarked while learning to process images by the use of mathematical morphology, was that not much is really known or written down about when a particular operation is applicable to a given situation. While much *is* known about the properties of each operation and its identities, there are few places in the literature which provide simple criteria for when a particular type of operation should be considered, or which tell when it is guaranteed to perform a certain desired task. Part of the problem here may be that there has not been sufficient mathematical attention given to these questions.

As an example of this, consider Figure 1.1 which is a simple binary image of some different kinds of tools. Suppose we see that the ones we want to extract (the Allen wrenches) are smaller in size than the ones we want to ignore. Morphological theory suggests that we should consider some type of 'opening' to discriminate them because openings address questions of size. That is fine once we know that size is the discriminating factor. But when exactly does it become appropriate to *think* about using an opening? What is the mathematical condition which must hold before *consideration* of an opening becomes a good idea, especially when we are dealing with not just

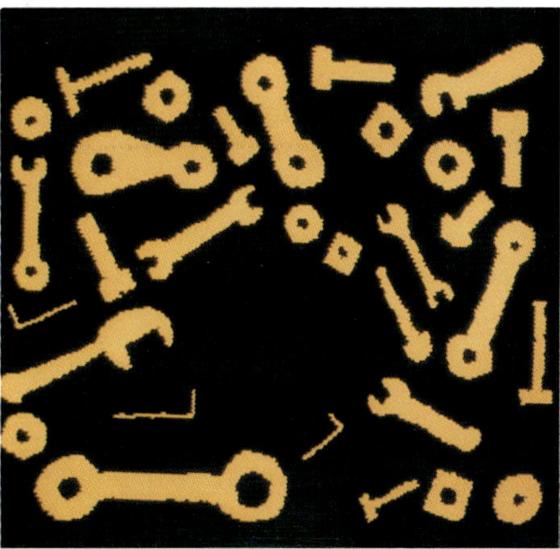

FIGURE 1.1. Binary Image of Tools (Courtesy ERIM)

one image, but tens or hundreds simultaneously? If we could answer this question, we might have a more concrete rule to give to novices for when to think about using an operation, and we would also have a test that we could put into computer form so that a machine could help us decide on the applicability of the operation, in addition to finding the distinguishing criteria for us automatically.

Going further we can ask, what is the condition which guarantees that a particular opening can *completely* discriminate the objects we want from the ones that we don't want, over *all* examples? If complete discrimination is not possible, how can we get a measure of the extent and the consistency with which a particular opening discriminates the two sets over a sample of images? How then can we automatically choose the opening parameter which *best* discriminates the two?

Very often, algorithms are developed by a trial and error process, with the operations being applied intuitively based on the experience of the researcher, using visual feedback to provide the measure of goodness for the result. The field thus remains a kind of 'black art', rather than a science. In addition, it presently requires months or years of experience before an individual develops a real grasp of these operations and knows when and how to use them.

While trial and error development may be good for learning how to develop algorithms, it is too slow and unreliable for a production style of development, because it places a large time and memory burden on the developer. Given that the operators we use are really just mathematical functions working on numbers or sets, it is not unreasonable to suppose that we can do better by using the computer more heavily in conducting tests and in choosing operators.

Likewise, when algorithm developers have to test the overall usefulness of an operation, or compare it against another alternative, often the only means available for doing so is through the visual feedback provided by the resulting images. While for simple cases, visual feedback may be quite adequate, in others, such as choosing between different possible smoothing operations, the eye is not such a reliable guide. Visual judgements can be very subjective, and inconsistent if the number of examples considered goes above five or ten. As the number of examples increases, more image memory is needed to store the results, and more flipping back and forth between them is required to make an overall assessment of the operation's utility. Especially when trying to compare three or four alternatives on each of several images, it begins to tax the memory of the developer to keep track of what happened in each case, and to make a valid judgement.

The development of formal measures of goodness, which would return measures or distributions of measures for the developer to evaluate, would do much to relieve this mental burden, and lead to more consistent, informed decisions. Larger samples could also be used, since the task of running the tests to get the measures could be given to the machine, leaving the algorithm developer to simply evaluate the results based on summary statistics.

A final related problem of algorithm development is that the search for optimal parameters is also commonly done by trial and error on small samples, simply because the mechanics of making the tests on larger samples is very cumbersome, and powerful measurement tools are not available which can do this kind of search automatically, returning to the user the optima and a summary of the variation present. This too is an area where giving some more of the work to the machine would help to improve the quality of algorithms and the speed of their development.

All of these considerations, and others, have contributed to my interest in working on the topic of the automatic generation of image processing algorithms for set recognition. In the next section I will attempt to formalize the class of problems this work is trying to address.

1.1 Problem Definition and Characterization

The subject of this book is the automatic generation of image processing algorithms of a particular type, namely those which 'recognize' certain objects or sets within an image, by the use of morphological processing tools. The problem to be solved in its simplest form can be expressed

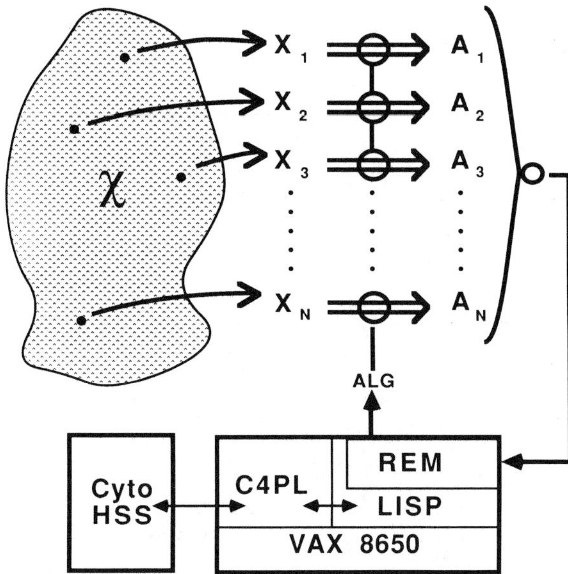

FIGURE 1.2. Abstract Representation of Problem

formally in the following way (see Figure 1.2):

> *Suppose I have a set of example images X_1 to X_N chosen from a sample space $\mathcal{X}$. For each example 'i', I have a result mask A_i, which represents the set of pixel locations I want to be selected from X_i. The question to be answered is, "What is the (minimal) algorithm (if any) which selects A_i from X_i (within a certain tolerance) for all 'i'?" The hope is that this algorithm will also select the correct set A_i from X_i, given any other X_i from the space $\mathcal{X}$, assuming that the sample used for training is sufficiently representative of this space.*

The problem statement above assumes that the algorithms are chosen from a language space which is made up of various combinations of a group of basic image-to-image operators.[1] Presumably too, the A's chosen for each X as the target set have something in common—they should consist of pixels belonging to the same type of objects in the different images, distinguishable from others based on distinct criteria such as size, length, distance, position, brightness, and so on. (The sets A_i in the sample are

[1]This assumption makes it impossible to generate a trivial algorithm which works only on the training set by simply storing each X_i/A_i pair into a look-up table for later comparison and retrieval.

typically chosen by a human observer using an interactive painting program or similar tool). Obviously, it is not hard to make the task impossible by choosing a set A randomly for each X; the real world difficulty of image processing is that even if the A sets are carefully chosen to be exactly the same type of object in each image, it is not always easy or even possible to find an algorithm which succeeds in choosing A_i and only A_i from each X_i in $\mathcal{X}$, much less a minimal one. *Our hope is that if the data are sufficiently regular, and if we are smart about how we search for an algorithm, and we are given a certain tolerance for error, then we can with some success formalize and efficiently automate the process of searching for a reasonable algorithm which performs this task.* The goal is to reduce the time it takes to find such algorithms, to formalize the application of operations and the methods for choosing them, and to develop automatic testing procedures for finding parameters to use with them.

I call the task described above a *set recognition* task, because what we are looking for are sets of pixels (A_i), within each example (X_i), and because we are in a sense trying to 'recognize' such sets based on common characteristics that they share, despite internal differences which they may have, and external variations of the context or background in which they sit. We will not be concerned here with other generic types of problems in image processing such as 'enhancement', 'filtering', 'segmentation', 'classification', or 'understanding'. The recognition problem differs from segmentation in that we are not dividing the image into regions and edges in order to produce a simplified image description—we simply want to find those regions which correspond to our objects of interest. In some ways it is like a classification problem of order 1, since we are trying to put pixels into two classes—those which belong to an object of interest or target, and those which do not. In general though, classification problems have more classes to discriminate, and they tend to use 'multi-band' data where each spatial location is measured over different spectra instead of the single band data which I will employ. They also generally use just intensity information in making the classification, rather than the spatial information that I will use here. Likewise, image understanding involves relating image constructs to real world scene descriptions, and goes beyond the scope of the recognition problem as described here, which operates only in terms of the image data.

Others might call this a 'concept learning' or 'pattern recognition' task, and indeed this work is closely related to work in both of those fields. It differs, however, because the goal is to select populations of locations (sets), rather than previously decomposed whole objects. In concept learning, for example, we usually have distinct objects given to us with known feature sets. The goal is to discover the relationship between these objects or the combination of features which constitutes the desired concept, and which allows us to select the appropriate whole objects (e. g., [Scott & Vogt 1983]). In statistical pattern recognition, we typically compute a set of features or measures on objects in an image, and then use linear combinations of these measures to select the objects which satisfy the concept. In general, we do not think in terms of individual pixels here, though one could extend the basic idea to individual pixels.[2] In structural or syntactic pattern recognition, again, we are usually concerned with global properties of primitive objects, rather than properties of individual pixels.

The set approach is characterized by the fact that the basic elements under study are pixel locations, and we 'recognize' the objects we want by using set operations to select the pixel populations which compose them. This selection is based on measures computed on the pixels, rather than on whole objects. We also 'combine' these measures, or rather the subsets which they represent, by non-linear operations such as union, intersection, and set complement, rather than by the linear combinations of features which are used in statistical pattern recognition.

1.2 Mathematical Morphology

The term 'morphological' in the book's title refers to the fact that the methods we will use and the algorithms we will generate are based on *mathematical morphology*, which is a set-theoretic approach to image processing developed primarily in Europe at the Ecole des Mines in Fontainebleau, France, by J. Serra, G. Matheron and others, with additional independent contributions by S. Sternberg, K. Preston, P. Maragos, and others in the United States. The choice of morphology as the basis for the present work was made for several reasons. First, it represents a fairly coherent, unified approach to image processing which is well-bounded. This stands in contrast to the image processing field in general, which includes a huge variety of techniques for which no overall theory has existed until recently.[3] Second, it is a non-linear approach, whose operations lose information at every step. In fact, the hallmark of the morphological approach is to *manage* the selective loss of information, so that the desired objects can be extracted, and the rest ignored. I felt it would be interesting to explore how to formalize the use of such non-linear tools in arriving at solutions to recognition problems, particularly since many of them preserve inclusion relationships, and this fact can be used to guide and to sub-divide the search for a solution.

Finally, mathematical morphology is an *algebraic* approach to image pro-

[2]In this case the set approach would still differ by using non-linear combinations of the features computed on each pixel.

[3]The situation has now begun to change with the work of Ritter & Wilson [1987], and Dougherty & Giardina [1987], both of which represent efforts to unify the field.

cessing; it is a kind of 'algebra of images' for mathematically studying shapes. Its algorithms are expressed in a common form as image equations which are concise, easy to read and manipulate, and which obey certain algebraic identities. These algebraic identities make it possible to speed up certain calculations or to reduce the amount of hardware needed to perform them. The ability to express algorithms algebraically is important for the task of automatically generating them using a symbolic representation as in Lisp, because a great deal of manipulation is required, which would be much more difficult if each operation were expressed in terms of individual pixels, rather than as whole images, or if the operators were from a command-oriented language, each with very different syntaxes and parameter sets.

There are four major routes or sources to use in trying to design a system for automatic generation of morphological recognition algorithms. The first is to study the image operators themselves—their definitions, properties, and identities. This kind of information is important not only for selecting operations for a given situation, but also for ordering them and for optimizing computational sequences. The second route is to study specific *criteria* which can be used to separate a desired set from its context—criteria such as size, distance, orientation, grey levels, adjacency, topology, orientation, position, and shape. These criteria are really what guide the search process for an algorithm. They have a loose connection with the operations in that there is usually more than one operation which can address a given criterion, and some operations speak to more than one criterion, especially when we add a structuring element. Thus although the relationship between criteria and operations is not exactly one to one, there is a certain amount of linkage which allows us to consider just a few possible types of operations, once we have focussed on a particular criterion. The most complete description of the operations and criteria in print to date is [Serra 1982]. Those of interest to this effort will be described in Chapter 2.

The third route to understanding image processing algorithms is to study actual examples or cases of problems which were solved in the past, and the sequence of operations used to solve them. This helps us to understand not only why certain operations were chosen for a given problem, but also why they were sequenced in a particular way. The close examination of such case studies is the best way to generate heuristics for how to order operations, and for deciding on search strategies to use in looking for algorithms. We will look at a few simplified examples in Chapter 3; I have studied many more of these in the course of my study of morphology and in my work experience of the past several years in France and in the United States. Indeed, one of the reasons I spent 2 1/2 years in France was to be able to be in close contact with Serra's group at the Ecole des Mines and their work, while implementing morphological programs and solving applications problems using these techniques in the area of medical imaging

[Vogt 1988c], for Thomson-CGR in Paris.

A final route for gaining an understanding of the image algorithm development process is to study the behavior of an 'expert' when faced with a new problem, by doing an analysis of protocols. This requires having an expert with whom to work, and the time needed to collect protocols and analyze them in order to obtain a coherent 'body of wisdom'. I did not have this kind of resource available to me for this project; however, in the course of my own learning about morphology and about how to solve problems on my own, I tried to pay special attention to the procedures I was going through, and to think about how these could be sped up or automated. It was this attempt to formalize the process I was using that led to the choice of the topic described here, and the work that has subsequently followed.

As a final note, I should mention that the overall search strategy and basic methodology to be used here are not entirely specific to mathematical morphology—it would indeed be possible to extend the same ideas to include a wider range of image transformations such as linear convolutions, correlations, and so on. I decided here to restrict the scope of the project to only certain morphological operations, partly because of their inclusion properties and also in order to limit the task to something manageable.

1.3 Related Work

In many ways, the REM system (to be described in the chapters which follow), is an experiment in automatic programming.[4] Automatic programming problems are typically expressed in terms of a problem domain, a target language, and a specification language.[5] The problem domain of the REM system is set-recognition image processing tasks which can be solved by simple algorithms using the operators of mathematical morphology. These operators are based on set-theoretic notions of union, intersection, and complement, rather than on linear and statistical notions. The target language used by REM is a directed graph and/or algebraic representation of image processing algorithms made up of these types of operators. Specification of problems is by *example*—that is, the user specifies the desired output image (representing the object or set of pixels to be 'recognized') for each input image, by drawing it—nothing is specified about how

[4]**REM** stands for '**R**econnaisseur **E**nsembliste **M**orphologique', or, in English, 'Morphological Set Recognizer'. **REM** is also borrowed as an acronym from psychology for '**R**apid **E**ye **M**ovements', a phenomenon which occurs during dreaming as images flash briefly across the screen of consciousness, akin to the manner in which REM flashes its calculations on the image display.

[5]See Rich & Waters [1988] for a recent review of this field.

to solve the problem, or what operators or strategy to use. (This would not preclude augmenting the system at a later date to accept suggestions or a partially structured algorithm from the user).

While the REM system can be thought of as an excercise in automatic programming, it differs in several important ways from the more classic general purpose programming systems in the literature such as PSI and CHI [Green 1977; Barstow 1977; Green and Barstow 1978], PECOS [Barstow 1979], and LIBRA [Kant 1979]. First, all of the data objects are of the same type—fixed size two-dimensional images. Second, all of the operators are image to image transforms. Some are unary (e. g., Complement), others are unary with non-image parameters (e. g., Erosion by a disk), and still others are binary combinations of two images (e. g., Union or Max). There are, however, no looping or branching constructs required—each operator is a self-contained unit which takes one or two input images and produces one output image. This makes the problem more homogeneous than that faced by other automatic programming systems.

Another way in which the REM system differs from other automatic programming systems is in the size of the operands. The images used in REM are made up of several thousand pixels, and real images generally have hundreds of thousands. The sheer size of these objects puts a premium on manipulating them—at every stage one has to try to reduce the number of calculations to be performed on images because this is what really takes up time. Measures or queries on images have to be done only when really necessary (even when using high speed image processing hardware), summary data must be saved to prevent later re-calculation, and wherever possible all questions which require the same image information should be asked at the same time. A further consequence of the large number of elements in each image operand is that the probability of obtaining algorithms which work perfectly on real images is almost zero. One is forced from the beginning to think about how to measure the goodness of algorithms in degrees, rather than to think in terms of proofs of correctness, as some automatic programming systems along the lines of DEDALUS [Manna & Waldinger 1978] have done. For this reason, although REM makes use of mathematical knowledge about how its operators work, it does not operate by trying to construct a proof that a given algorithm computes exactly the desired image function.

REM follows what I would call a 'strict' definition of automatic programming, in that it performs a search for an algorithm within a fairly large space of possible ones, constrained by whatever it can learn about the example data it is given. As such, it has to contend with the typical problems of search programs—i. e., how to avoid searching the whole space without risking to miss the solution, and how to perform its search in a fairly intelligent manner. This differs from what I would call automatic programming in the 'loose' sense, in which one is simply taking a high-level program specification, with perhaps some details left unspecified or to be filled in by context, and translating it into a lower level executable version. The latter capability is essentially demonstrated in any kind of high-level language compiler, though some systems have gone beyond this in the sense that they accept natural language input, or only partially specified programs [e. g., SAFE—Balzer, Goldman, & Wile 1977, or NLPQ—Heidorn 1976]. Because the primary task of these systems is one of translating and filling in details, rather than search, I would separate them from 'strict' automatic programming systems like REM and the others mentioned above which have to look for solutions.

There has been very little written in the literature as yet on the subject of the automatic development of image processing algorithms. One paper by Sakaue and Tamura [1985] describes an automatic programming system (in the loose sense) which simplifies the use of the SPIDER image processing software package, by allowing users to specify program schemata without all of the normal parameters. By checking the linkages between different program modules, it can either determine what the missing parameters should be automatically or query the user for any others. This system does not do any problem analysis or search for an algorithm. Another paper by Haas [1987] creates programs to measure important features of binary objects or configurations of them. The objects are modeled using context-free attribute grammars, and this, combined with hardware for doing binary template matching, are used to recognize the objects, before making the goal measurements. Once again, this type of system does more of a translation of a high-level program into a lower-level one, using the object grammar descriptions in the process, than it does a true search. It is restricted to recognizing only examples of objects that have previously been modeled for it in the grammar, and for which primitive templates have been constructed.

The only system which attempts to at least in part generate morphological algorithms automatically is that of Gillies' [1985] dissertation, also carried out at the University of Michigan. In this case the problem was to distinguish whole images drawn from different populations, where each image contained one instance of a class of objects to be discriminated (e. g., samples of different scripts, cartoon characters, or classes of abstract shapes). Each image contained only the binary object instance, reasonably centered, and surrounded by empty space with no extraneous clutter. The goal was to find an algorithm that left a non-Nil result when applied to image instances of a desired class, and Nil results when applied to all other classes. (Multiple algorithms of this type would be used for a multi-class problem).

In contrast, the system proposed here has to distinguish pixels in the *same* image (desired vs. undesired), where all the images are drawn from the same problem population, and where the images may have various types of noise or clutter present in them. In both systems the problem is to find a single algorithm that works for all instances, but whereas the REM system has to recognize sets of *pixels* based on specific criteria, Gillies' system is really designed to classify whole images as coming from one image population or another.

The two systems also differ significantly in how they generate algorithms. In Gillies' system, each algorithm is considered to have two 'parts'—the structural part which consists of the overall algorithm form, the base operators to use, their combinations, and their sequencing, while the second or 'parameter' portion consists of the set of specific structuring element shapes to be used by each base morphology operator. In Gillies' system the basic structure or form of the algorithm was chosen by a human, as a function of the classes of images to be discriminated. The structuring element parameters were chosen by a semi-random process.[6] The work demonstrated that the genetic search was able to find good discriminating structuring elements faster than a completely random search could.

The REM program is based on a different philosophy and differs significantly in that it is entirely automatic—all of the work is done by the machine using formal and heuristic methods. There are no random elements, and no parts of the algorithm are pre-supplied by a human operator. The system uses expert knowledge about the operators and their effects to propose, test, and finally select the ones to be used. Humans provide only the masks representing the desired results. Sequences are built up in a step-by-step fashion of progressive refinement, again based on formal understanding of the underlying operators and their effects. While at present the problems solvable by the system are limited in difficulty, this stems from the fact that our formal understanding of the process of algorithm development is also presently quite limited. As this understanding progresses, the REM system would be expected to be able to solve longer and more complex problems.

1.4 Goals and Objectives

The major goals of my research into the area of automatic generation of image processing algorithms have been four-fold. The most important goal has been to understand and to be able to describe the image algorithm development process well enough to formalize it. After that, the second goal has been to develop a system that could automatically generate recognition algorithms from a set of examples and 'correct-answer' masks, for simple problems at first and later on for more realistic ones. In addition to making it possible to generate such algorithms automatically, I have wanted also

[6] A *genetic algorithm* [Holland 1975].

to improve our understanding of the operations and their uses, by formally stating the conditions for which each is applicable and for which each is guaranteed to work. Finally, I have wanted to help speed up the algorithm development process by creating automatic analysis tools for this purpose, which test the utility of operations for a given problem, and which choose parameters for them automatically, on the basis of looking at a complete set of examples at once.

The implementation of all of these elements represents an enormous task, requiring special image processing and measurement hardware for image manipulation speed, successfully interfaced to AI hardware for control of the search, the storage of domain specific knowledge, and for data analysis, decision making, and algorithm representation. The work covered by the current project was only enough to illustrate the overall concept and the major principles involved. Stated formally, the specific objectives for this project were the following:

1. To define the problem domain of the system to be implemented for automatically generating set recognition algorithms based on mathematical morphology.

2. To describe the theory behind this activity, including a model of the algorithm development process, and arguments that the search process proposed is equivalent to an exhaustive search, in that it will not miss any correct solutions.

3. To implement a system for the selected problem domain, and describe its components and operation.

4. To present results showing that the implemented system is indeed capable of solving set recognition problems within its domain automatically, and that it does so in a manner significantly faster than what could be obtained by an exhaustive search.

The major programming activity of this work has been in designing and implementing a system to solve set recognition problems, and in demonstrating the principles of this design on a number of simplified image problems, using a small set of operations, and images which are smaller and less complex than those which are normally used. The implementation effort included several important subtasks, which are listed below:

- Design of languages for representing problems, as well as for representing partial and complete algorithms.

- Establishment of evaluation mechanisms to be able to measure how well an operation solves a given subproblem, and to measure the difficulty of the subproblem before applying any operators.

- Construction of expert knowledge for each set of operations, to know, mathematically, when they might be worth considering, as well as when they will definitely work.

- Writing of procedures to collect data for each class of operator, and design of data structures and a methodology for managing this collection of data about operators in the course of testing their utility for given subproblems.

- Development of strategies for the analysis of operator data—testing for different alternative solutions, looking for the easiest ones first, saving the more difficult ones as fallbacks.

- Development of techniques for the generation of new algorithm code, once the analysis suggests that a given operator will work.

- Establishment of strategies for subdividing problems, and for choosing the best solutions from a list of alternatives with their corresponding benefits and costs.

- Development of techniques for building up complete algorithmic solutions by successive program transformations, and for deciding which lines of transformations to pursue at a given time, based on the state of the overall search tree.

All of these elements had to be implemented in some form, and work together in synchrony, before the overall system could be expected to solve any problems, or even produce any output.

I believe that the work described here will be a very important step toward the development of systems which automatically generate image processing programs, either independently or with the aid of a human for guiding the search. *It should be possible to reduce weeks or months of algorithm development time by a researcher to hours or perhaps days of computation.* The key elements to all of this are the formalization of how to decide when to try the operations, of how to decide when they work, and of how to decide when to decompose a problem into separate parts. There has been almost no work done on any of these topics to date—image processing continues to be more of an art than a science, guided primarily by intuition based on the experience of the individual, and passed down from expert to apprentice by prototypic examples. I think that the attempt to formalize the use of these operations will not only give us better and faster algorithm development systems in the future, but will also improve our understanding of the operations themselves in the present, and will promote the development of more powerful analysis tools for doing this kind of work. In the near term it should help to improve the quality of results and the turn-around time needed to solve problems in this domain—something that is essential if the use of computerized vision is to grow rather than stagnate.

1.5 Organization of the Book

The material which follows is divided into four major parts. The first, (Chapter 2) discusses the problem domain, mathematical morphology; the second, (Chapter 3) is concerned with issues of theory regarding the recognition of sets; the third, (Chapter 4) describes the different parts of the REM system; and the fourth (Chapter 5) presents results obtained by running the program on a large number of different example problems.

In a bit greater detail, Chapter 2 begins with a review of mathematical morphology, describing its approach to image processing, the types and properties of its operations, their identities, and the criteria used to distinguish sets of pixels. In particular the major operations that will be of interest here will be described in detail, including their properties and the criteria which they address. The concept of 'band' operations is also introduced, something which is central to the operation of the REM system and the basis of the target language it uses for constructing algorithms.

Chapter 3 discusses the theory needed for solving set recognition problems formally—the data structures required and the different elements of the search process. The chapter begins by presenting a few simple examples to illustrate the types of problems we want to be able to solve, the effects of some of the operations, and the mathematical representation of the algorithms. This chapter also presents arguments that the search mechanism used cannot miss any correct solutions within its defined repertoire or algorithm space; i. e., it is equivalent in power to an exhaustive search over that space.

In Chapter 4 all of the major parts of the REM program are described in detail. The first section covers the four levels of languages used in the system to represent and execute algorithms or algorithm fragments. Section 2 presents a top-level view of the REM program. Section 3 describes the overall search process. Section 4 and its subsections describe the different elements of the problem solver—the evaluation mechanism, the 'expert' system which proposes operators to examine, the data collection system which produces distributions for these operators on each of the images being tested, the analysis system which studies the distributions generated, looking for possible solutions which it then proposes as algorithm code modifications, and finally the portions for choosing between different proposed solutions and for dispatching those chosen. The last section of Chapter 4 describes the system output and display, which are used to illustrate and monitor the image processing operations and other activities of the REM system as it executes.

Chapter 5 describes the current capabilities of the system and presents results obtained by running the REM program on more than 50 different types of example problems. An analysis of the speed of the system is also presented here to show that in addition to being complete, the search process used is also significantly faster than an exhaustive one.

Finally, Chapter 6 presents my conclusions regarding the present work, the major contributions made here, and suggestions for further extensions and improvements, toward the goal of having practical systems of this type in the future.

2

Review of Mathematical Morphology

We begin our study of the automatic generation of set recognition algorithms by giving a brief review of the essential elements of mathematical morphology. Though we cannot hope to cover all of it in such a short space, I will describe the main principles of the theory and give a general overview of the major classes of operations, while concentrating on those which made up the focus of this project. The purpose of this review is to provide the mathematical underpinnings for what is to follow. Some of the material in this chapter, particularly that in Sections 2.3, 2.5, 2.7, 2.10 and 2.11, is treated in greater depth in Serra [1982].

2.1 Types of Image Data

The fundamental units in the field of image processing are, of course, images. Images come in several different types, depending on the data elements, or *pixels*, which compose them. The breakdown is similar to the data or variable types defined in statistics, which is not surprising, since statistical methods are among the most widely used for analyzing image data.

First there are the *binary* images composed of pixels with only two values: on/off or 1/0, similar to a yes/no variable in statistics. Next are the state or *phase* images, which correspond to categorical variables in statistics. Each element of these images has one of 'K' possible states or labels (categories) which serve simply to identify their type. As the categories are simply names or labels (i. e., nominal data), there is no implied relationship between them. A 'null' category (label 0) is sometimes used to mark pixels which do not belong to one of the categories of interest. A binary image can be thought of in one sense as a phase image with only two such states.

The next image type is the *grey-level* type, for which each pixel has one of K (usually 2^N) values, which are strictly *ordered* (i. e., ordinal data in statistics). These values may be thought of as the integers 0 to $K-1$, or as the rationals $(0/K-1)=0$ to $(K-1/K-1)=1$. A binary image can be thought of in these representations as a specialized case of grey-level image which has only the extreme values of 0 and $K-1$, or 0 and 1, respectively.

Alternative schemes can be used for representing signed ordinal data.

Sometimes additional assumptions can be made with grey-level data, beyond the one of strict ordering. If the actual 'distance' between adjacent steps in the range is known to be the same, then the data is said to be *metric*. The metric steps may be additive or multiplicative. An example where this would not hold would be with a sensor that had uneven or variable response across its operating range—the grey values would still be ordered, but they wouldn't represent metric steps. Likewise, in some cases the response of a sensor is carefully calibrated so that a given grey value always corresponds to the same physical measurement, even in different images. In this case, the data are *absolute* instead of *relative*. (It is clearly possible to have absolute data which does not have the metric property, and metric data on a sliding, non-absolute scale). These additional properties of grey-level data are important because they affect what types of operations are valid to use on the images. Too often they are simply ignored.

A fourth type of imagery, called *multi-band* or *multi-spectral*, is used when independent sensor data is available for each spatial point in an image. For example, a color image has three bands, one for each of the red, green, and blue channels. Each band by itself is a grey level image but they are considered as a single unit since the values of a pixel in each band all come from the same spatial location. In some cases, such as with satellite imagery, one can have as many as a dozen bands in one image.

Finally, in some applications where the range of pixel values must vary over several orders of magnitude and a log representation of the data does not provide the desired sensitivity, each pixel is represented by a floating point number, giving us a *real* image. The values may range from 0 to 1, 0 to $+\infty$, or $-\infty$ to $+\infty$. In cases where complex numbers are needed to represent the phenomena under study, two real bands are used.

In this project, I have considered only binary images and grey level images representing data which is metric and absolute.

2.2 Images as Sets in Space

An image, in general terms, is always defined over some (usually discrete) space. In our case it will be a two-dimensional rectangular array, but the number of dimensions may vary—one could have a linear binary or grey image of one dimension (i. e., a signal), or a three or more dimensional one. A three dimensional binary image defines the locations of a solid object in 3-space; a grey level 3D image might give the density of each location in the space; and a multi-band 3D grey image might contain both density and temperature information for each location. We will not be concerned with these more general notions of 'images' here, but it is important in any case to draw a distinction between two different kinds of dimensions—those which make up the value(s) or data type(s) of each pixel, and those which make up the spatial support of the image. These are often confused—particularly when one speaks of a 2D grey level image as a '3D image', with the grey value representing the third dimension. In fact what we have in this case is an image with two *spatial* dimensions and one *value* dimension which holds grey level data.

The spatial support of an image is what we mean when we refer to the dimensionality of an image. While for practical image processing it must be composed of discrete elements, for theoretical purposes it need not be—indeed, most of the mathematical theory developed in France at the Ecole des Mines' Centre de Morphologie Mathematique, has been based on the Euclidean spatial supports, R^n. The discrete space realizations of the theory are treated in this framework as degraded versions of Euclidean space phenomena which occur as a result of the digitalization process.

Discrete spatial supports generally appear in the form of a uniform grid which divides the space into equal-sized units. For one dimension, there is only one possible grid—a linear sequence of blocks. In two dimensions, however, the grid may be either a square raster array, or a hexagonal honeycomb array. In three dimensions there are four possible grids which can be used [Serra 1982, p. 196]. While the hexagonal grid offers certain advantages over the square one in two dimensions, particularly with respect to connectivity questions, it is a little more difficult to represent and to work with. In this book, I will consider only square grid images.

Mathematical morphology is a set-theoretic approach to image processing and analysis, which considers images to be sets in the underlying support space, and manipulates them using set-based operators like union and intersection. Originally developed for binary imagery (for which set operations are about the only ones available), it has since then been extended to grey-level and other types of images. Here we will be interested in how it applies to both grey-level and binary images. The key difference between these two with respect to morphology concerns how they are combined—binary images are combined by the usual set operations of union and intersection, but for grey images these two concepts are extended to mean the max and min functions, respectively. In this sense the grey level images of morphology are related to the theory of fuzzy sets [Goetcharian 1980]. A binary image is an exact set which has only the values 0 and 1, while a grey image is a fuzzy set with pixels that take on values between these two.

Morphology differs from fuzzy set theory and from purely statistical approaches to image processing in its geometric component—that is, in the way in which individual elements of the support grid are related to each other spatially. Unlike purely statistical approaches, which ignore spatial relationships and information about shape, the morphological approach to image analysis has been constructed precisely to answer questions about shape in a rigorous mathematical way. Its hallmark is to make use of specifically shaped probes, called *structuring elements*, or simply *stelts*, to extract different kinds of shape information from images. In the following sections of this chapter we will see how this is done.

2.3 Operation Classes and Properties

Before we look at morphological image operations in detail we need to define some terminology relating to the general classes that image operations fall into, and the types of algebraic properties which they can have.

Basically there are six generic classes of image operations. The *Generation* class are those which create a new image of a given type from scratch through parameterized graphic operations. The *Combination* class consists of those which take two images of a given type and produce a new image of the same type, by somehow combining each pair of elements from the input images into a new element (e. g., add, subtract, min and max). The *Transformation* class takes one input image of a given type and transforms it into another image of the same type. Here we distinguish between point transformations which depend only on the value of an individual pixel, and neighborhood transformations which also depend on some set of (not necessarily nearest) neighbors of each pixel. The fourth class is the *Measurement* class which takes in an image of a given type and reduces it to a number or distribution representing a measurement on the image. Locations or addresses of image features are also included in this class. Next, the *Conversion* class includes those operations which take an image of a given type and convert them to a different image type (e. g., thresholding). A sixth class, *Display*, involves operations whose sole purpose is to control how an image is presented to a human—we will not be concerned with this class here.

Another thing which distinguishes mathematical morphology from other approaches to image processing is that it is an *algebraic* approach. Its algorithms can be written in algebraic form, and most of the morphological image operations have algebraic properties which govern their behavior. Before discussing individual operations, it is important to know the names and definitions of these properties. Here we will talk in the abstract, using names like Ψ_1 and Ψ_2 to indicate different operations, and X_1 and X_2 to indicate operands. Table 2.1 lists the most important of these properties.

The **Commutative** and **Associative** properties allow us to rearrange the order of certain expressions without modifying the resulting computation. The fact that order is not important in certain expressions also

Commutative:

$$(X_1 \,\Psi\, X_2) = (X_2 \,\Psi\, X_1) \tag{2.1}$$

Associative:

$$((X_1 \,\Psi\, X_2) \,\Psi\, X_3) = (X_1 \,\Psi\, (X_2 \,\Psi\, X_3)) \tag{2.2}$$

Distributive:

$$(X_1 \,\Psi_1\, (X_2 \,\Psi_2\, X_3)) = ((X_1 \,\Psi_1\, X_2) \,\Psi_2\, (X_1 \,\Psi_1\, X_3)) \tag{2.3}$$

Extensive:

$$(X_1 \,\Psi) \supseteq X_1 \tag{2.4}$$

Anti-extensive:

$$X_1 \supseteq (X_1 \,\Psi) \tag{2.5}$$

Increasing:

$$X_1 \supseteq X_2 \Longrightarrow (X_1 \,\Psi) \supseteq (X_2 \,\Psi) \tag{2.6}$$

Monotonic Increasing:

$$S_2 \supseteq S_1 \Longrightarrow (X \,\Psi\, S_2) \supseteq (X \,\Psi\, S_1) \tag{2.7}$$

Monotonic Decreasing:

$$S_2 \supseteq S_1 \Longrightarrow (X \,\Psi\, S_1) \supseteq (X \,\Psi\, S_2) \tag{2.8}$$

Idempotent:

$$((X_1 \,\Psi) \,\Psi) = (X_1 \,\Psi) \tag{2.9}$$

Homotopy Preserving:

$$HT(X_1) = HT(X_1 \,\Psi) \tag{2.10}$$

Duality of Operations:

$$(X \,\Psi_1) = (X^c \,\Psi_2)^c \tag{2.11}$$

TABLE 2.1. Important Operator Properties

reduces the amount of complexity we would otherwise have to deal with if this were not true. The **Distributive** property allows us to simplify certain expressions and thus to compute more complex operations by combining simpler ones.

The **Extensive** and **Anti-extensive** properties give a certain directionality to the operations which have them. If an operation is extensive, it always gives us more than what we had at the beginning. Likewise, if it is anti-extensive, it always gives us less. One use of these properties is in taking differences, since it means that we can arrange for our differences to always be positive, and so that what is left always comes from either the inside or the outside of the set from which we started, but not both. In addition, if we are searching for an operation which will give us a certain result which *contains* the set we have at present, we can restrict our search to just the extensive ones, and vice versa. Finally, we note that if an operation is extensive with respect to the foreground of an image, then it is anti-extensive with respect to the background, and vice versa; this relates to the duality property discussed below.

The **Increasing** property insures that the behavior of an operation is not wildly erratic—specifically, that it *preserves containments.* If this is not true, analysis of the behavior is more difficult, since it means that one could start with an image Y that contains X, and get a result out $Y_1 = (Y \ \Psi)$ that does not necessarily contain $X_1 = (X \ \Psi)$. If the containment property varies as X and Y change their size or shape, it suggests that the operator is not really a local one—it must somehow be using global information about the sets. Useful operations which do not have the increasing property include the 'border' operation, which removes all of the interior pixels from a binary shape, and other 'difference' operations. All of these, however, are based on operations which *do* have the property. Most of the basic operations defined in morphology have the increasing property, because it is needed for proofs of their behavior, especially in the limit with Euclidean support spaces. The only commonly used operation which has the opposite (decreasing) property, is the complement operation which reverses the background and foreground of an image, i. e., $(X_1 \supseteq X_2) \Longrightarrow ({X_2}^c \supseteq {X_1}^c)$.

The **Monotonic Increasing** and **Monotonic Decreasing** properties refer to how operations behave with respect to parameters that they use (in this case structuring element shapes, described in Section 2.9 below), when those parameters satisfy a relationship of containment. For a monotonic increasing operation, if we apply a series of increasingly larger structuring elements to an original image, we are guaranteed to get a series of increasingly larger results. For a monotonic decreasing operation, we will get a decreasing sequence of containments in the results. These two properties are important for certain proofs and especially for classifying pixels based on a series of similar-shaped structuring elements which vary only in size. Because these two properties are so important for what follows later on in

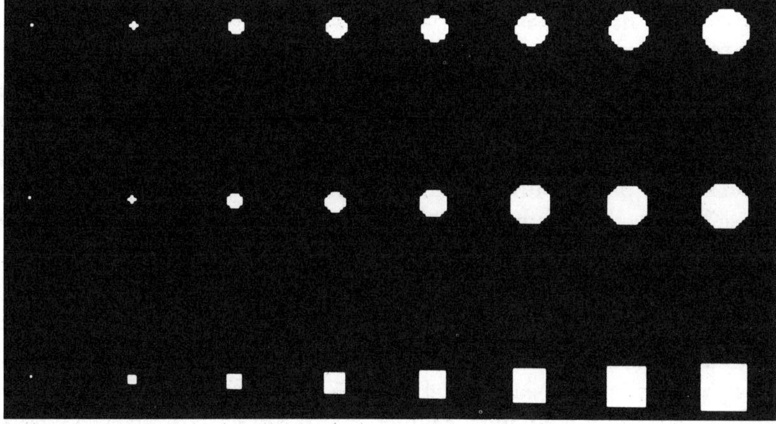

FIGURE 2.1. Stelt Sequences for Disks, Octagons and Squares

the implementation, I want to emphasize them by digressing for a moment to explain them in more detail.

Consider the notion of a structuring element sequence:

$$\mathcal{S} = \{S_0, S_1, S_2, \ldots, S_N\}$$

In particular, let us assume that $\mathcal{S}$ is an *increasing sequence* in the sense that:

$$S_0 \subseteq S_1 \subseteq S_2 \subseteq \ldots \subseteq S_N \quad \text{or} \quad \forall_{i \geq 0},\ S_i \subseteq S_{i+1}$$

In general, such sequences will be composed of objects with the same basic shape (such as disks, octagons, hexagons, squares or lines), centered on the origin, but with different sizes as indicated by the index 'i'. Figure 2.1 illustrates sequences of disks, octagons and squares. In each case S_0 is just the origin, which is in fact an identity element for all of the basic morphology operations. What we would like to be able to do for such a sequence is to classify each pixel in a binary image by associating it with just one of the S_i's in $\mathcal{S}$, as a function of the base operator (erosion, dilation, opening, closing) we choose, and the spatial characteristics of the pixel itself. How can we do this?

The requirement for being able to unambiguously classify each pixel in a binary image using a morphological operator and an increasing sequence of structuring elements, is precisely that the set of results we obtain by applying the operator with each of the structuring elements satisfies a strict containment relationship. In this case we say that the operator is *monotonic.* Specifically, we say that the operator Ψ is *monotonic increasing* if the result for each larger element *contains* the previous result, or *mono-*

tonic decreasing if the result for each larger element is *contained by* the previous result.

Monotonic increasing operations must be extensive while monotonic decreasing operations must be anti-extensive. Given the monotonic increasing property with an extensive operation, we can construct a classification for each pixel in the background of a binary image, by associating it with the size of the structuring element in the sequence for which it *first* appeared in the sequence of the results. By projecting out only the class of each pixel in the image, we get a distribution much like a histogram, although based not on the pixel's brightness, but rather on its spatial characteristics.

Likewise, given the monotonic decreasing property with an anti-extensive operation, we can classify each pixel in the foreground by associating it with the size of structuring element for which it *last* appeared in the sequence of results, and we can construct a distribution from this classification. The distributions so-generated give us tools or measures for distinguishing groups of pixels based on their spatial characteristics. These will become the basis of the analysis for all of the binary operators implemented in the REM system. (This is described further in Chapter 4).

Continuing, the **Idempotent** property means that for some operations, performing them once is the same as performing them any number of times. This keeps us from having to decide how many times to apply them, and it also means that such operations provide convenient 'stopping points' in an algorithm sequence—they serve to demarcate the meaningful units or separate parts of a total algorithm. When this property is combined with the extensive or anti-extensive properties it also allows us to simplify certain expressions or sequences of calculations.

The word '**Homotopy**' refers to the connectivity relationships which exist in an image (in a sense its 'digital topology'). The homotopy tree (HT) is a structure which expresses these connectivity relationships in a simplified way (see Serra [1982], p. 89). The homotopy preserving property simply states that an operation does not modify these relationships even though it may simplify the image in other ways. The medial axis transform is an example of such an operation.

The **Duality** property is one which exists between two operations. It says that we can get the equivalent of a certain operation by performing the dual operation on the complement image, and taking the complement of the result. In another sense, when we perform the first operation on the foreground, we are at the same time performing the second (dual) operation on the background, and vice versa. The first sense allows us to use hardware that can perform only one of the operation pair, and use the complement operation with it to get the second; the second sense tells us that we don't have to concern ourselves about whether to use an operation on the background—it is sufficient to always focus on the foreground, and to consider the effect and utility of each of the dual operations upon it. Erosion and dilation are examples of morphological operations which are dual with respect to each other. Because of the fact that most morphological operations are either extensive or anti-extensive, it turns out that they are usually defined in dual *pairs*, one that does something anti-extensive to the foreground, and the other which does the same thing to the background that, when expressed in terms of the foreground, is an extensive operation.

The importance of these eleven properties will become more evident when we start to define specific operations in the following sections. The purpose of presenting them here has simply been to introduce and define them, so we can avoid doing this again later.

2.4 Criteria

From the morphological viewpoint, the process of trying to recognize an object in an image is really one of finding a way to discriminate the set of pixels which compose it from the others in the image space. The characteristics we use to perform this discrimination are called *criteria* (e. g., length, width, distance, etc.). It is not enough to look just for criteria which are true for all pixels belonging to the object—one also has to find characteristics which are, at the same time, *not* true of the rest of the pixels in the image. In trying to find criteria to discriminate these two sets of pixels, we begin with somewhat subjective visual criteria and eventually try to translate these into objectively measureable ones.

The set of criteria used to discriminate desired pixels from undesired ones constitutes a set of *assumptions* about the 'space' of possible images which are being examined—assumptions about the objects of interest themselves, and about the context in which they sit. Sometimes these assumptions can be derived from prior knowledge about the inspection task, the nature of the object, or the imaging device. In other cases they have to be derived empirically, by examining many samples of images and making measurements on them. In our case we are assuming that no prior knowledge is available, but it would not be hard to extend the idea to include prior knowledge—this would simply narrow the search space by reducing the set of alternatives to be examined.

The art of the algorithm developer is to first find such discriminating criteria visually by careful observation of the class of images under examination, and by studying the problem domain and the acquisition method used to generate the images. He must then use this information in conjunction with his knowledge of image operation behavior to generate a sequence of steps which will perform the discrimination accurately, despite the normal variations in the images of departure. *It is the knowledge of the behavior of image operations, and their relationship to visually perceived criteria, which constitutes the expertise of the image algorithm developer.* In general, this knowledge is fairly intuitive—there is very little in the literature of morphology which describes the precise conditions for which a particular operation is well-suited. One of the objectives of this chapter and this book, is to help clarify this situation for certain operators by stating exactly the discrimination which each one performs, as it is defined below.

There are many different types of image criteria that can be used either singly or in combination to discriminate sets. Table 2.2 lists some of the most important ones, organized into general groupings. The ones of most interest to the implementation effort that will described later are marked with an asterix. The purpose of going through these at this point is to introduce them so that we can refer back to them as we discuss the effects of each image operation in the sections which follow. Here again we are going to restrict our discussion to the image types of interest, i. e., binary and grey level images. Most of these criteria have been originally defined for binary images and were later extended to grey level ones; for some, such as size, those extensions are fairly straighforward, but for others, such as connectivity, they are more complicated and will not be dealt with here.

The first group, concerning grey level images only, includes criteria based on the absolute levels, on the changes in levels for pixels that are next to one another (such as high contrast edges), and on the relative levels within a locality compared to its surroundings (local peaks or valleys in the grey level surface).

The second group concerns questions of position, orientation, and direction. Position may be either absolute (addresses in the image space) or relative (a distance and direction vector from something else). Likewise orientation criteria may be either absolute or relative to the orientation of something else. Direction includes specifications such as 'left-of' or 'above-right-of'.

The third group includes the spatial characteristics of size, distance, spacing, and containment. The first three are fairly straightforward; containment refers to whether an object of a particular size and shape can fit inside another one. Different notions of size include maximum width, maximum length, area, and radius of circumscribing circle. Distance and spacing differ in that a distance measure shows how far away a pixel is from something else, while spacing refers to the size of the space to which it belongs.

Additional spatial characteristics of shape, curvature, and topology are listed in the fourth group. Shape is a very complex notion which can be described many ways, including some which are closely related to size. Topology here refers to things like the presence and number of holes, and the end points, triple points, and segments found in the medial axis of a binary image.

- Absolute Grey Level (Brightness)*
- Change in Grey Level (Contrast)
- Local Relative Grey Level

- Position*
- Orientation
- Direction

- Size*
- Distance*
- Spacing*
- Containment*

- Shape*
- Curvature
- Topology

- Adjacency
- Connectedness*
- Particle Measurements*

- Pattern
- Texture
- Noise Characteristics
- Periodicity
- Symmetry

TABLE 2.2. Common Criteria for Discriminating Image Elements

The next group of criteria in the table refer to relationships between, and measurements of, individual connected components. Adjacency expresses the relationship 'next to'; connectedness includes some of the topology issues mentioned above as well as how different parts of an image are connected to each other—whether a particle or hole touches the outer frame of the image or not is also included here. If we consider the individual connected components of a binary image one by one (perhaps by labeling each with a different value), we can conduct measurements on the particles themselves which may allow us to choose the ones we want (based on perimeter, area, compactness, orientation, etc.—clearly there is overlap here with the types of measures we can make on entire images).

Finally, the last group includes criteria related to specific patterns, texture, noise characteristics, and periodicities and symmetries found in images.

The distinction between different types of visual criteria is not always hard and fast, as is apparent from looking at this list. The point here is simply to illustrate some of the kinds of 'themes' that image algorithm developers can use in devising recognition algorithms. In general, for each type of criterion, there are only a few operations which address it, so the search space is fairly small once they have been identified (it remains only to select the best parameters). The more serious difficulty is in finding criteria to begin with that are sufficiently invariant across the space of images to be usable in constructing a robust discrimination algorithm.

Before closing this section, I want to identify the image criteria that will be of specific interest to the implementation effort to be described later on. *Absolute brightness* will be the only criterion used to distinguish pixels in grey level images. For binary images, a number of different criteria will be used, starting with *absolute position* in the image along the X and/or Y dimensions. *Distance* from the foreground or background will be used—in fact several different notions of distance including isotropic distance, octagonal distance, chessboard or 8-connected distance, and city-block or 4-connected distance. In addition the *size* class of foreground and background pixels will be considered, which is defined as the largest-sized object of a given shape to which the pixel belongs, such that the complete object fits entirely within the foreground or background, respectively. We will consider four types of shapes in these size measurements—disks (isotropic size), octagons, squares (8-connected), and diamonds (4-connected). In addition, we will also be using criteria based on *connectedness*, which in effect, put all pixels belonging to the same particle or hole in the same class, based on some characteristic of the entire unit. The specific criteria will include size of the particle or hole (four different types, similar to those described above for classing individual pixels based on size); overall internal length of the particle or hole, measured by the 8 and 4-connected skeletons; and finally smoothness/roughness of the particle or hole, measured both with respect to disks and to squares. (This criterion can also say something about the object's shape).

When added together, this makes for about 35 different criteria that we are using to possibly distinguish pixels. (This does not include all of the different possible parameter settings that can be associated with each one, which would push the total to many thousands of different specific criteria). One essential point to mention here is that although the example problems the REM program has solved are particular instances using particular images based on a given criterion, the *method* the system uses is general. If we were to give the system *any* other set of images such that the pixels of interest could be distinguished on the basis of the same criterion, it would solve that problem just as well—its answers are not dependent on the specific input images, but rather on the *relationships* between the pixels in those images. More will be said on this subject in Chapter 3.

Now that we have finished describing the notions of image types, their support spaces and treatment as sets, the broad classes of operations and the properties they can have, and the types of criteria which can be used to discriminate sets and to construct algorithms, we are ready to look at how some of the basic image operations are defined, their effects, and their specific meanings in terms of the criteria which they address. We will concentrate primarily on the binary operations, since these are simpler to describe and to illustrate, and are the focus of the implementation effort.

2.5 Set Operations

Since mathematical morphology is a set-theoretic approach to image processing, the operations upon which it is based, and the ones which it uses to combine two images (sets) into one, are the set operations of union, intersection, complement, set subtraction or intersection complement, union complement, and symmetric difference. We will define each of these in turn. Figure 2.2 illustrates some of the binary set operations which we will examine, while Figure 2.3 illustrates their grey level counterparts.

The *Union* operator, (X_1 Union X_2) or ($X_1 \cup X_2$) is $\{p : p \in X_1 \vee p \in X_2\}$ for binary images. It is commutative and associative. For grey level images the union is interpreted as the *Maximum* of the input images, taken point by point, which also has the same properties. In terms of the 'meaning' of an algorithm, the union of two binary images or sets is used to find the *disjunct* of the criteria represented by each of the binary sets. For example, if one of them represents 'all pixels which had a grey value greater than 128' and the other represents 'all pixels which were more than a given distance away from some object' the union of these two binary images would be a direct representation of the set which satisfied the disjunct of these two criteria.

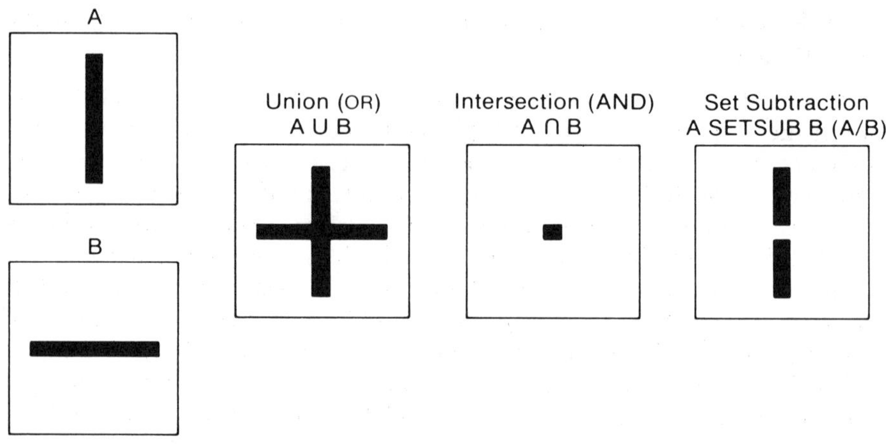

FIGURE 2.2. Binary Set Combination Operations (Courtesy ERIM)

For grey level images, the interpretation of the union as maximum is used to combine different types of high structures in the grey image topography that have been selected by independent techniques. It can also be used to select the pixel value from all of several input images which satisfies some criterion to the greatest extent, again as expressed by its grey value.

The *Intersection* operator, (X_1 Intersect X_2) or ($X_1 \cap X_2$) is $\{p : p \in X_1 \wedge p \in X_2\}$ for binary images; it too is commutative and associative. For grey level images it is interpreted as the *Minimum* of the input images, and has the same properties. The binary intersection is used in the set approach to image processing to express the *conjunct* of the meanings represented by the individual input sets, in the same way that the union expresses the disjunct. The grey level intersection or minimum is generally used to put together low structures which have been selected independently, or to select each pixel value from whichever image satsifies some criterion to the least extent.

The *Complement* operation, X^c, negates the image, which in binary swaps all zeros with ones and in grey takes the maximum value complement, ($63 - v$ in the case of REM) which amounts to flipping the image or grey level surface 'upside-down'. The meaning of the complement is of course to express the negation of the expression represented by the input image. In general, it is not used alone very often—it is most often used in conjunction with the intersection or union to obtain residues of the basic morphological operations. Alternatively it can be used to perform morphological operations on the image 'background'.

There are two operations which make use of the complement and the

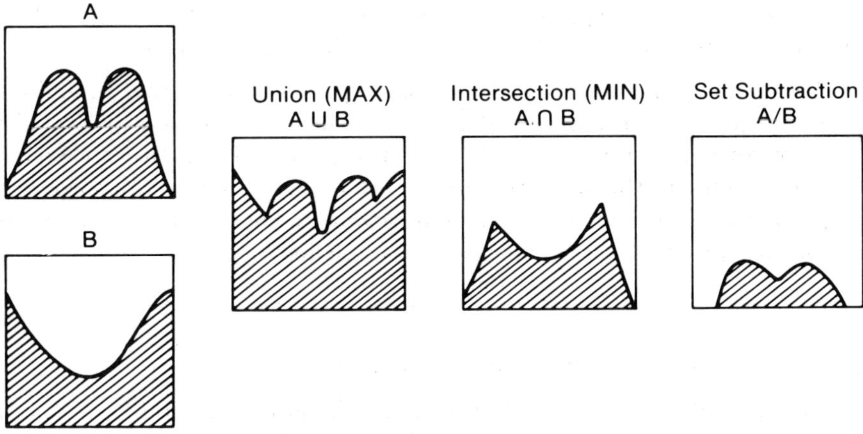

FIGURE 2.3. Grey Level Set Combination Operations

union or intersection in order to obtain the residues of sets (see Section 2.12). The first, *Intersection Complement* or *Set Subtraction* is defined as $(X_1 \cap X_2{}^c)$ or $(X_1 \setminus X_2)$. It is used to remove, from X_1, the part of X_2 which is also part of X_1. (For anti-extensive residue operations X_2 is contained in X_1). For binary images the interpretation of this is straightforward, as all pixels which are in X_1 and not in X_2, or all of those which satisfy the expression represented by X_1 but not those which also satisfy X_2. In grey, things are a little less clear—there are actually three definitions which could make sense and which are consistent with the binary case. They are:

$$\min(X_1, X_2{}^c)$$

$$\min(X_1, [X_1 > X_2])$$

$$\max(0, X_1 - X_2)$$

The first of these just involves substituting the grey level definitions of intersection and complement into the binary set subtraction definition. This would seem to be a reasonable approach but the result is an operation that looks bizarre—it returns the minimum of the distance of X_1 from 0, and of X_2 from the maximum value—which has no clear utility or meaning. The second definition returns the original value of X_1 wherever it is greater than X_2, else 0. This is useful in some situations but less so than the third definition, the 'clipped subtraction', which returns $X_1 - X_2$ unless the result is negative, in which case it returns 0. This third definition is the standard one used for grey level set subtraction, corresponding to the binary intersection complement. It can be interpreted as the extent to which the image expression represented by X_1 is brighter or higher or truer than that represented by X_2, and we will write it here as: $X_1 \uplus X_2$.

The *Union Complement* is defined as $(X_1 \cup X_2{}^c)$. It is used to add on to the set X_1, all of the pixels in the image which are not part of X_2. (For extensive residue operations X_1 is contained in X_2). Again the interpretation or meaning of this operation for binary images is straightforward. For grey images the correct definition is analogous to that for the intersection complement, namely the 'clipped addition', defined as:

$$\min(\text{Maxvalue}, X_1 + X_2)$$

This operator will be written here as: $X_1 \Cup X_2$.

The *symmetric difference* operator, $(X_1 \text{ Symdiff } X_2)$ or $(X_1 \parallel X_2)$ is expressed in binary as all of the pixels which are in X_1 or X_2, but not both; i. e., $\{p : p \in X_1 \parallel p \in X_2\}$. This is rarely used except to measure the extent to which two sets are equivalent or not. In grey morphology it represents the same thing as the absolute difference of the two arguments.

2.6 Positional Masking and Thresholding Operations

The next group of operations we will consider are some positional masking ones, which select pixels in an image based on their addresses, and some conversion ones which allow us to create binary images either from scratch or by converting grey images to binary.

(Zero) generates a null image (or null set $\emptyset$) which covers the support space; (One) generates an image for which all pixels have the maximum value; this is also known as the universal set $\mho$. These two generator operations are included mainly for completeness; they are used primarily as part of predicate tests for images which are 'empty' or 'complete', but $\mho$ is also necessary for defining the positional masking operators which appear next.

The positional masking operators for binary images which we will be using in the implementation are defined as follows:

$$\begin{aligned}
(\text{X Infmask-X A}) &= (X \perp^x A) = (X \cap \mho_{X(A)}) \\
(\text{X Infmask-Y A}) &= (X \perp^y A) = (X \cap \mho_{Y(A)}) \\
(\text{X Supmask-X A}) &= (X \top_x A) = (X \cup \mho_{-X(MaxA+1-A)}) \\
(\text{X Supmask-Y A}) &= (X \top_y A) = (X \cup \mho_{-Y(MaxA+1-A)})
\end{aligned}$$

Here $\mho_{X(A)}$ refers to a translation (see next Section) of the universal set in the X direction by A pixels—this will cause zeros to appear on the left side, all the way up to column $A - 1$. The infmask operations thus remove (set to 0) all pixels whose X or Y address is less than A. They are anti-extensive operations since they always return a set contained in the original. The supmask operations on the other hand add (set to 1) all pixels whose X or Y address is less than A; they are extensive operations. In another sense the infmask operations maintain all of the 1 pixels whose absolute position in the image is greater than or equal to address A along the dimension designated, while the supmask operations maintain all of the 0 pixels in the same region. By combining these basic positional masking operations with the complement, we can easily generate others which select areas of the image whose address is *less* than or equal to a given address, or between two address limits.

The importance of the infmask and supmask operations is to implement windowing and to restrict the area of interest to rectangular subsets of the support space, by changing the value of all the other pixels to another state, 0 or 1, depending on whether we are really interested in the particles or the holes (foreground or background) of the original binary image. These operators also work as defined for masking grey level images, but we will not make use of this aspect of them here.

We turn next to the thresholding operation which is written:

$$\text{(G Threshold T)} \;=\; (G|_T^+)$$

This operation converts a grey image to a binary one, by changing all pixels $\geq T$ to the maximum value (e. g., 255 or 1). This operation addresses the criterion of absolute grey level or height, i. e., 'All pixels $\geq T$'. By combining this operation with the complement, we can also produce operations which express 'All pixels $< T$', or within a certain range $T_1 \rightarrow T_2$. If grey values represent the extent to which something is true, then the threshold operations can also be interpreted in terms of selecting pixels which satisfy 'degrees of truth' criteria.

Relational expressions between grey level images such as:

$$[G_1 \{<, \leq, =, \neq, \geq, >\} G_2]$$

are expressed in the above manner, where any one of the six relational operators listed may be specified. These expressions, like the threshold, also generate a binary image, but which represents those locations where the pixels in G_1 bear the appropriate relationship to the corresponding ones in G_2. Again the meaning of the relationship may be thought of either in terms of brightness values, height, or degrees of truth, depending on the interpretation of the grey images used as input. The importance of the relational expressions is that they provide a way to generate an image which, when considered as the embodiment of an algebraic set expression, represents a *comparison* between the two grey level or fuzzy algebraic set expressions which were represented by the input images.

2.7 Translation-Based Morphology Operations

The next class of morphology operations we want to look at are transformations (one input image, one output image) based on the use of the set operations described above with translations of an image. These form the most basic and important operations of morphology. We begin with translation.

(X Translate p) or X_p is a translation of the image set X by the point or translation vector p=(x y). Elements of X which move outside the image window are lost, while outside elements which move into the window are usually given a default value of 0 or, in certain cases, the maximum possible one.

The next step is to take a group of such translations of X, and combine them with a union or intersection. If S is a set of translation vectors $(s_1, \ldots, s_N)$, the Minkowski addition is defined as the union of translations of X by all of the vectors in S, that is:

$$\text{(X Minkadd S) or } (X \oplus S) = \bigcup_{\forall s \in S} X_s$$

and the Minkowski subtraction is:

$$\text{(X Minksub S) or } (X \ominus S) = \bigcap_{\forall s \in S} X_s$$

These names are in deference to Minkowski, who first defined the first of them. Since for grey images, union means maximum and intersection means minimum, the corresponding definitions for grey images are:

$$MAX_{(\forall s \in S)} X_s$$

$$MIN_{(\forall s \in S)} X_s$$

In what follows, we will assume that the reader understands this difference in expression for grey images and will simply speak in terms of union and intersection. The same algebraic results hold for both types of images.

The group of translation vectors making up S is called a *structuring element* (or *stelt* for short), and may be represented as simply a list of vectors, or as an image with an origin or center specified—each point representing a translation vector in the structuring element. The image representation is better for understanding how the stelt interacts with the image; the list representation is better for implementing the corresponding operation. (It requires less space). The essential difference between an image and a stelt is that a stelt must have an origin, while an image need not have. The fact that a stelt can be thought of as an image, however, is important because it allows us to build larger ones from smaller ones, through the same operations that we use for images, as will be shown subsequently.

The Minkowski addition is an increasing operation, which is extensive (contains the original) whenever the origin belongs to the structuring element. (See [Serra 1982] for proofs). It does not preserve homotopy nor is it idempotent. Likewise the Minkowski subtraction is increasing, and anti-extensive whenever the origin is part of the structuring element. These two operations are dual, because $\forall_X \forall_S \ (X \oplus S) = (X^c \ominus S)^c$, in both the binary and grey level image cases. Another property of interest they have concerns inclusion of structuring elements, namely:

$$S_2 \supseteq S_1 \Longrightarrow (X \oplus S_2) \supseteq (X \oplus S_1)$$

$$S_2 \supseteq S_1 \Longrightarrow (X \ominus S_1) \supseteq (X \ominus S_2)$$

In other words, Minkowski addition increases monotonically with increasingly larger structuring elements, while Minkowski subtraction decreases monotonically. The proofs arise from the fact that the union, on which Minkadd is based, is an extensive operation viewed with respect to the first operand, while the intersection is anti-extensive under the same interpretation. These properties will be important when we look at sequences of structuring elements with the same underlying shape.

In terms of an image, $(X \oplus S)$ returns all of the locations which can be 'reached' by placing the origin of S at all of the points of X, while $(X \ominus S)$ returns all of those points which are covered by *every* translation of X by the vectors of S.

The Minkadd operation also has several additional important properties. First of all, like the union, it is both commutative and associative, while neither of these holds for the Minkowski subtraction:

$$(X \oplus S_1) = (S_1 \oplus X) \quad \text{(True)}$$

$$(X \ominus S_1) = (S_1 \ominus X) \quad \text{(False)}$$

$$((X \oplus S_1) \oplus S_2) = (X \oplus (S_1 \oplus S_2)) \quad \text{(True)}$$

$$((X \ominus S_1) \ominus S_2) = (X \ominus (S_1 \ominus S_2)) \quad \text{(False)}$$

Note that we are interchanging the role of S_1 here; sometimes we think of it as an image, sometimes as a structuring element. As mentioned above, this is perfectly legal, and it allows us to construct new stelts from other ones by the use of the union, intersection, Minkadd and Minksub operations. In general, only the union and Minkadd are used for this purpose, since the others do not preserve identities, and would usually increase rather than decrease the amount of computation required. The real importance of the commutativity and associativity of the Minkadd operation is with respect to structuring elements, rather than images. That is, we are not usually interested in $(S \oplus X)$, but it is important that:

$$(X \oplus (S_1 \oplus S_2)) = (X \oplus (S_2 \oplus S_1))$$

$$(X \oplus (S_1 \cup S_2)) = (X \oplus (S_2 \cup S_1))$$

$$(X \oplus ((S_1 \oplus S_2) \oplus S_3)) = (X \oplus (S_1 \oplus (S_2 \oplus S_3)))$$

$$(X \oplus ((S_1 \cup S_2) \cup S_3)) = (X \oplus (S_1 \cup (S_2 \cup S_3)))$$

This means that we don't have to worry about the order in which we apply the structuring elements, as long as they are joined by unions or Minkadds. It also means that a long sequence of stelts, joined together by $\oplus$, represents a *single* structuring element, which is, in a sense, the geometric *product* of those stelts (the order invariant combination of factors). We can enumerate the structuring elements constructable by the Minkadd operation in terms of *combinations* of smaller elements, rather than permutations. The construction of a large stelt by the union of smaller ones, on the other hand, can be considered as taking a geometric *sum* of them. As with integers, while there are few ways to get a given product, there are usually many, many ways to get the same sum. Likewise, we can obtain large integers (or stelts) faster by taking products, than by taking sums.[1]

There are two important operations which are used only with structuring elements, the *transpose*, S', and the *reflection*, $S^\wedge$. The transpose flips each point in the stelt across the origin of its *spatial* support (i. e., $x \rightarrow -x, y \rightarrow -y$, for 2D stelts). The reflection negates the *value* dimension of a stelt, in a sense turning it 'upside-down' ($z \rightarrow -z$). The latter has no effect on 'flat' or binary structuring elements whose 'value' at each location is zero. It only modifies grey level stelts which are not flat—we will not be too concerned with these here. Note also that the transpose has no effect on stelts which are already symmetric about the origin. For completeness we

---

[1] As an aside, I would mention that it is also possible to define the concept of 'prime' stelts, analogous to prime numbers, as those which cannot be further decomposed by Minkowski addition. Unlike the integers, however, there is no unique prime factorization theroem for structuring elements, because some of them can be constructed in multiple ways. For example, an elementary hexagon can be constructed from two triangles or three segments, all of which would be considered 'prime'.

should point out here that:

$$(S_1 \oplus S_2)' = (S_1' \oplus S_2')$$

and

$$(S_1 \oplus S_2)^\wedge = (S_1^\wedge \oplus S_2^\wedge)$$

The importance of the transpose and reflection is that they allow us to define two new operations, erosion and dilation, which have very important practical meanings in terms of what they do to images. Dilation is defined as:

$$(\text{X Dilate S}) = (X \circledvee S) = (X \oplus S') = \bigcup_{\forall s \in S} X_{s'}$$

Erosion is:

$$(\text{X Erode S}) = (X \circledwedge S) = (X \ominus S'^\wedge) = \bigcap_{\forall s \in S} X_{s'^\wedge}$$

These operations have the same properties as the Minkadd and Minksub operations, including the duality relation, but their meanings are slightly different. In terms of the original set X, the dilation returns all of the locations where the origin of S can be placed such that it has a non-empty intersection with X, that is:

$$(X \circledvee S) = \{p \mid X^c \not\supseteq S_p\} = \{p \mid S_p \Uparrow X\} = \{p \mid S_p \cap X \neq \emptyset\}$$

This meaning of the dilation is valuable for certain probability calculations since the area or volume of the result is a direct measure of the probability that the object S placed at random in the space will intersect the set X, and that tells us something important about X. In another sense the dilation leaves as background only those locations where S fits completely in the background, so it tells us something about the 'containment' of S in the background.

Likewise the erosion tells us about containment in the foreground, since $(X \circledwedge S)$ returns all of the locations where we can place the origin of S such that S completely fits inside of X. Alternatively, the area or volume left after the erosion expresses the probability that a shape S placed at random in the space is completely contained within X:

$$(X \circledwedge S) = \{p \mid X \supseteq S_p\}$$

The Minkowski addition and subtraction by S do not have such clear probabalistic interpretations in terms of S and X, which is why the erosion and dilation are more commonly used. In any case, erosion equals Minkowski subtraction and dilation equals Minkowski addition for structuring elements which are symmetric about the origin, and that is the case in the large majority of practical situations. The importance of defining the Minkadd and Minksub operations is that they express the simple concepts of union and intersection of translations on which all of the morphological operations are based, and because the Minkadd is useful in building larger structuring elements from smaller ones. In what follows we will concentrate on the terms erosion and dilation, and assume that the reader understands their relationship to the Minkowski operations.

2.8 Structuring Element Decomposition

In this section, we want to look closely at how structuring elements can be decomposed into smaller, simpler elements. This is important because it is the basis for assessing the image processing costs of the morphology operations which use them. The motivation for decomposing stelts is based on the following four extremely important identities, two for dilation/Minkowski addition, and two for erosion/Minkowski subtraction:

Union Decomposition:

$$(X \oplus (S_1 \cup S_2)) = ((X \oplus S_1) \cup (X \oplus S_2)) \tag{2.12}$$

$$(X \ominus (S_1 \cup S_2)) = ((X \ominus S_1) \cap (X \ominus S_2)) \tag{2.13}$$

Dilation Decomposition:

$$(X \oplus (S_1 \oplus S_2)) = ((X \oplus S_1) \oplus S_2) \tag{2.14}$$

$$(X \ominus (S_1 \oplus S_2)) = ((X \ominus S_1) \ominus S_2) \tag{2.15}$$

The first two are called the 'union decomposition' identities while the second two are called the 'dilation decomposition' identities. The first two allow us to decompose a Minkadd or Minksub operation by any large structuring element into a union of Minkadds or intersection of Minksubs by smaller stelts whose union makes up the larger one. This 'union decomposition' is akin to breaking a large number into a 'sum' of smaller components. It gives us the ability to do complicated operations with limited hardware, if we are willing to wait (the time cost is proportional to the area of the large stelt). Note that the first equation is just the distributive property for dilation with respect to union, and that the second is similar but with a slight twist—that we have to change the union to an intersection when we do the distributing for erosion.

The second pair of equations allow us to decompose the dilation or erosion by certain large structuring elements (those which can be constructed by the Minkowski addition of smaller ones), into a sequence of successive dilations or erosions, each step operating on the results of the last one.

Not all structuring elements of interest can be constructed in this way, but many can. This again allows us to do more complicated operations with limited hardware; it also allows us to speed up the calculation of certain operations, because dilating the small stelts together has a 'multiplicative' effect—the time it takes to construct the large stelt is now related to the sum of the small stelt areas, rather than its own, larger area. These identities make it possible to do certain operations which would otherwise be too time consuming if performed as a 'sum', and thus have an extreme practical importance.

The practical importance of the dilation decompositions has been exploited by cellular image processing machines such as the ERIM Cytocomputer, which basically performs morphological operations on 3x3 windows and concatenates these into a pipeline to construct the equivalent of using larger stelts, based on the dilation decomposition identities. The limitations of this approach occur mainly with large, round structuring elements such as disks, because the 3x3 window limits the number of slant angles that can be produced, so that the best possible approximations to disk stelts have 16 sides. In addition, there are other shapes which are not compact, that also cannot be decomposed into 3x3 neighborhoods.

There is a third type of decomposition which can be used in some situations, called 'recursive' decomposition, and it is based on the following observation. Suppose I dilate together a group of structuring elements, all of which have only two points each, and one of these points is always the origin. The fact that one of the points is the origin means that at each step, I am simply keeping the previous result as is, and taking the union of it after translation by the second point:

$$X \oplus (\mathcal{O} \cup p) = X \cup X_p$$

and thus

$$\begin{gathered} X \oplus ((\mathcal{O} \cup p_1) \oplus (\mathcal{O} \cup p_2)) = \\ (X \oplus (\mathcal{O} \cup p_1)) \oplus (\mathcal{O} \cup p_2) = \\ (X \cup X_{p_1}) \oplus (\mathcal{O} \cup p_2) = \\ (X \cup X_{p_1}) \cup (X \cup X_{p_1})_{p_2} \end{gathered}$$

If the dilation of these two point structuring elements gives a stelt that I want to dilate with X, then I can perform the equivalent operation by recursively taking the union of each result with a translation of itself, and making *that* the result of the next step. In other words, $R_0 = X$, and $R_n = R_{n-1} \cup (R_{n-1})_{p_n}$, for all of the points p in the original two point stelts which were not origins. This recursive generation of a structuring element has a kind of 'exponential', rather than a simple 'product' effect, because it builds upon itself. Using this technique it is possible to generate structuring elements extremely quickly, but only those which can be expressed as the dilation of two point stelts. This class includes all digitally convex, symmetric objects, as well as 'regular' lattices or groupings of convex symmetric objects. It requires hardware which can perform arbitrary translations. The MVI Image Flow Computer was based on the recursive decomposition technique [Abbot et al, 1988]. We write the recursive dilation and erosion operators as follows:

Recursive Dilation: $(X \overset{r}{\oplus} S)$

Recursive Erosion: $(X \overset{r}{\ominus} S)$

Two advantages of recursive decomposition of stelts are that it allows one to generate 'perfect' digital disks[2] and other round, symmetric convex shapes very quickly, and that it lets one generate line segments in log base 2 of their length steps. The latter is done by shifting by one and taking the union, then shifting the result by two, taking the union again, shifting this by four and eight and so on until the desired length is obtained. Here we get the sense in which this method works 'exponentially'. Cellular systems with 3x3 windows, in contrast, need a number of steps equal to about half the length of the line.

The disadvantage of the recursive technique is that it only works for a limited class of stelts, and it can be slower than the cellular approach for stelts which are small. It should also be mentioned here that the idea of a structuring element as a small image breaks down when we look at the set of translation vectors in a recursive decomposition because it is quite possible that some of them will be duplicates. While the order of application of them is unimportant (commutativity), and associativity still holds, the distribution over union does not hold—we have to replace it by the following equations, which use concatenation of the vector lists in place of union:

$$X \overset{r}{\oplus} (S_1 \mid S_2) = (X \overset{r}{\oplus} S_1) \overset{r}{\oplus} S_2$$

$$X \overset{r}{\ominus} (S_1 \mid S_2) = (X \overset{r}{\ominus} S_1) \overset{r}{\ominus} S_2$$

The subject of structuring element decomposition is a vast and interesting one, which has only begun to be studied carefully in the past five to ten years. It is still not known what is the class of objects which can be generated by 3x3 decompositions, the generation of grey level stelts is still a fairly complex problem,[3] and the problems of finding the shortest decomposition or best approximate decomposition to a desired shape have not

[2]That is, the best possible digital approximations to true Euclidean disks.
[3]See [Shih and Mitchell, 1987 and 1988] for some work on this subject.

yet begun to be solved. More complex schemes such as combining unions of convex symmetric primitives to obtain a desired shape are also possible, but have not been explored to any great extent. The practical concern of this topic is: What operations can be performed on images within a reasonable time, or, what is the time or hardware cost of individual algorithm steps, so that we can evaluate complete algorithms in terms of their speed versus accuracy? We are going to leave these major unsolved problems now, for other researchers to work on, and turn next to an examination of some of the major commonly used structuring elements, including how they are used, how costly they are to compute, and what are their other advantages or disadvantages.

2.9 Common Structuring Elements

In mathematical morphology, structuring elements are considered to be probes or special tools for looking at images. We choose a particular one, let it interact with the image, see what has changed, and because of what we know about that interaction, we can deduce something about the shapes in the original image. The art of the morphologist is to to know which operation/stelt pair to select in order to extract information about a given aspect or criterion of interest. Over the years, a library of the most commonly used probes has grown up, each of which addresses different kinds of needs. The ones that will be of interest to us here include flat disks, pseudodisks, squares and hexagons (homothetics), lines, vectors, point pairs, single points, and rings. We will also mention some other special binary patterns used for connectivity operations, and a few of the non-flat grey level elements such as cones and half-spheres. Some of the most common stelts are illustrated in Figure 2.4.

The simplest structuring element is a single point. If the point is directly on the origin then we get an identity transformation or exact copy of the original; if not, the shape of the original set will be maintained but it may be translated and/or shifted in the grey level dimension. As a measurement tool, the point stelt is associated with the area measure for binary images, or the grey mass measure (the sum of all the grey values) for grey level images.

The next simplest element is the pair of points. Generally either one of the points is at the origin and the other is at some radius and angle away, or the origin may be centered between them (these two results will only differ by a translation). The first of them is used to focus on periodicities in an image as a function of angle and distance; the second is more useful for finding axes of symmetry. Because the space between the two points is empty, this type of stelt tells us where things vary together, independent

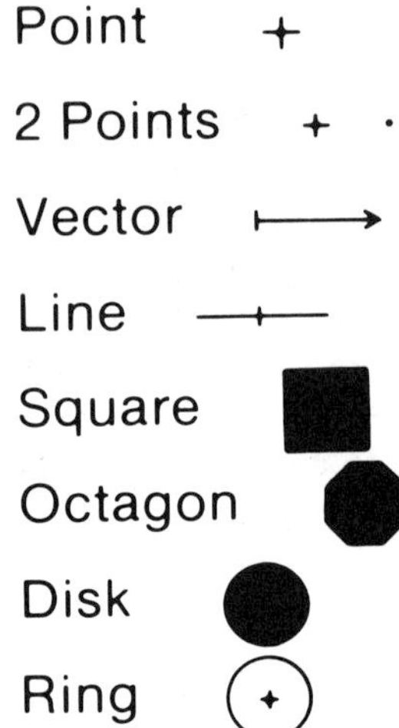

FIGURE 2.4. Common Structuring Elements

of what goes on in the middle. As such, it is used to measure the set *covariance*, which is the probability that a point p has a neighbor of a given type, (r, θ) or (Δ_x, Δ_y) away.

The *line* structuring element consists of a contiguous linear sequence of points, with the origin at the middle. (Again, since translation commutes with Minkadd and Minksub, placing the origin somewhere else would only modify the result by a translation). The line has a radius and an orientation, and is clearly used to locate linear structures, particularly those with fixed orientations, less than a certain length. If we want to find linear structures of unknown orientation, we can use a union of lines at different orientations. Line stelts can be used for the chord length distribution measure, which calculates the probability that chords of different lengths can fit in the original set. In some cases the origin of a line is placed at one end of it; in this case we call it a *vector*. Vectors are used when we are concerned about linear distance in a certain direction, for example, if we want to find something to the 'east' (right) and within distance 'd' of a known object. Line and vector stelts have a cost equal to the radius, R, if a 3x3 dilation decomposition is used, or about $\log_2 R$ for a recursive decomposition.

The term *homothetics* in general refers to shapes which differ only by a scaling factor; here we will restrict its definition to those shapes which have the same shape as an elementary neighborhood of the spatial support, that is, squares and hexagons, depending on the grid being used. Homothetics in this sense are special because they can be generated very quickly, by repeatedly dilating the elementary neighborhood by itself, on a cellular machine (cost equal to the radius), or by dilating together the recursive decomposition of one-half the edges, with a cost of about $N/2 \log_2 L$, where

N is the number of sides of the object and L is the length of each side. Compare this to the area generated, and we find that homothetics provide the greatest area for the least number of operations of all the stelts. They are used to get rough measures of size and distance, when accuracy needs to be sacrificed for speed. Here the hexagonal grid has an advantage over the square one, since a hexagon is much closer in shape to a disk than a square. Homothetics are also of direct value when objects of interest in the image have this shape. The homothetics allow us to get at the idea of 'cellular' or 'pixel' distance in an image, as opposed to the Euclidean distance idea which is expressed by the disk, described below.

The diamond or rhombus shape is another homothetic which is only a little more difficult to generate than the square, and which may be preferable in some situations. (It cannot be completely decomposed recursively, however). It allows us to quantify the notion of 'city block' distance in an image. By doing a little more work, a *pseudodisk* can be generated, which might correspond to an octagon or hexadecagon in the square grid, or to a dodecagon in the hexagonal grid. This gives us a shape that is a bit closer to a true disk than the corresponding homothetic, but at a cost which is not too much greater. For a cellular machine, the number of steps is actually about the same as the homothetic, but they are not all identical so the construction is slightly more difficult; for a translation-based machine, the cost is about twice that of the square. The pseudodisk can be viewed as a compromise between accuracy and speed, when the real desire is to know isotropic distance or size. Like the square, however, the pseudodisk can also be used when the object of interest itself is of this shape.

For questions of accurate (almost Euclidean) distance or size, the required structuring element is the digital disk. Unfortunately, 3x3 cellular machines can do no better than generate 16-sided pseudodisks. For the most common, smaller radii disks, this is not a problem, but for larger ones it is. Translation-based machines make it possible to generate perfect digital disks, at a cost of between one and two times the radius, depending on the exact decomposition, but the construction is fairly complicated.

The fact that digital disks are complicated to generate becomes even more of a problem when we want to do operations involving whole series of stelts with similar shapes—whether disks, squares, hexagons or lines.[4] The lines and homothetics have the property that a larger one (radius R) can be generated from a smaller one (radius r) simply by dilating the smaller one by certain additional elements to complete its 'Minkowski sum'. That is,

$$S_R = S_r \oplus S_{R-r}$$

[4]The use of stelt series which satisfy an inclusion relationship is important for classifying pixels and for choosing operator parameters.

and so

$$X \oplus S_R = (X \oplus S_r) \oplus S_{R-r}$$

and

$$X \ominus S_R = (X \ominus S_r) \ominus S_{R-r}$$

We can thus generate the result of a larger dilation or erosion by starting with a previous intermediate result which used a smaller stelt of the same type. (We save some steps this way). This can give us a fast technique for classifying pixels based on approximate distance from the background, for example, by using hexagons of successively larger radii. It does not hold true for the digital disks, however, because they don't satisfy the first equation above—we are forced to compute each element in a sequence of disk dilations starting from the original image X, so we are doubly penalized for using accurate disks in these cases. Digital disks also create problems when one is trying to generate measures or distributions for openings and closings—this will be covered in Section 2.11.

Another kind of structuring element which has been recently introduced into the field is the digital ring, and its generalizations to other types of closed loop shapes. The digital ring is just composed of the points on the boundry of a digital disk—it is a closed loop in the shape of a circle. Whereas the solid stelts like the squares and disks tell us something about the 'internal' size of things depending on whether and where a particular shape fits inside, the ring and other closed loops can tell us about the 'external' or 'circumferential' size of an object by indicating whether it can fit completely inside the loop or not. This could, for example, be used to distinguish long skinny objects from small compact ones. By combining a solid stelt probe with a closed loop one that fits around it, we can define a 'hit or miss' transformation ($X \circledast (S,T)$) which sets inner and outer limits on the patterns which it will accept [Serra 1982]. These transforms make up a very general class of pattern recognizers [Crimmins and Brown 1983]. Unfortunately, the cost of the ring and closed loop stelts is about equal to their length, since there is really no way to decompose them efficiently. [5] Approximations of fewer points are sometimes used to reduce this cost.

So far I have not spoken about grey level structuring elements, but for the most part nothing changes. All of the 'flat' structuring elements described above work exactly the same in grey as in binary, except that the unions and intersections of translations are replaced by Min's and Max's. The meanings and uses of the elements is also the same. Sometimes there is a temptation to give a 'third' dimension to such elements in the grey case, and to talk about 'cylinders' instead of disks, and 'walls' instead of lines, but this is really unecessary—there is no difference between the flat binary

---

[5] It is sadly *not* the case that: $X \oplus (S_1 \setminus S_2) = (X \oplus S_1) \setminus (X \oplus S_2)$.

disk and the flat grey disk. If we force the issue and add a height parameter to these stelts to give them a third dimension, we find that we gain nothing because the result is the same as using a flat stelt, and shifting either the original image beforehand, or the result image after the flat morphological operation (i. e., Z-shifts commute with the flat morphological operations). The only time things change is when the Z-surface of the stelt is not flat—for example if it has the shape of a half-sphere, a cone, a pyramid, or a paraboloid. Then we have to perform Z-shifts for each stelt point, as well as X-Y shifts, before taking the Min or Max. The meanings of the operations are still much the same for these true grey level stelts—i. e., where they fit in the foreground or background—but they give us some additional flexibility over the flat ones regarding what is accepted or rejected. True grey level stelts are fairly costly to generate, are used fairly rarely, and their uses are for the most part still not fully appreciated. As such, we will not deal explicitly with them here except to mention them for the sake of completeness.

One final class of structuring elements really involves stelt pairs, where one element of the pair specifies a condition on the foreground, and the other specifies a condition on the background, as in the case of the hit or miss transform described above. The most common of these involve just nearest neighbor elements, and are used to compute convex hulls of objects, and thinning transforms which reduce the number of pixels in a shape while preserving its homotopy. Typically, these two stelt patterns have to be *rotated* through all configurations of the elementary neighborhood to get an isotropic result, and then they have to be repeated again and again until the image stops changing. With a cellular machine, all rotations of the patterns can be tested in one step, and the cost is then about equal to the radius of the maximum disk that can fit in the original image (i. e., it is data-dependent).

The list of possible structuring elements is practically endless, but the ones we have discussed here are the most commonly used. For some applications where one is trying to recognize a specific shape that is known in advance, it may be worthwhile to design a new one. Except for this, though, it is often difficult to tell when a particular esoteric one would be very useful. For our own purposes, we are just going to concentrate on the disk, the octagon, the square, and the diamond. We will not cover use of the ring, the true grey level stelts, the point pairs, the four 45 degree lines, the eight 45 degree vectors nor the two element patterns used for pattern recognition and homotopy-preserving operations.

FIGURE 2.5. Original Binary Image

2.10 Erosions and Dilations

To recap a bit at this point, we have defined what we mean by images, saw how they are treated as sets, discussed the different classes and properties of operations and the kinds of criteria they can address. We then looked at some threshold and position mask operations for binary and grey level images, defined the set operations for combining images of these types, and then began to look at the basic operations of morphology, and how they are defined in terms of translation and the set operations. We discussed the properties of these base operations, defined the concept of 'structuring element', and showed how they can be constructed or decomposed and how this relates to the cost of using them. Finally we looked at some of the common types of structuring elements, and discussed how they are used. In the sections which follow, we want to look at some of the derived operations of mathematical morphology—to define them, and discuss their properties and meanings. We will begin with a quick review of erosion and dilation, which were mentioned briefly earlier.

Erosion:

$$(X \circledwedge S) = \bigcap_{\forall s \in S} X_{s'^{\wedge}}$$

Dilation:

$$(X \circledvee S) = \bigcup_{\forall s \in S} X_{s'}$$

Erosion and dilation are dual, increasing, non-idempotent operations.

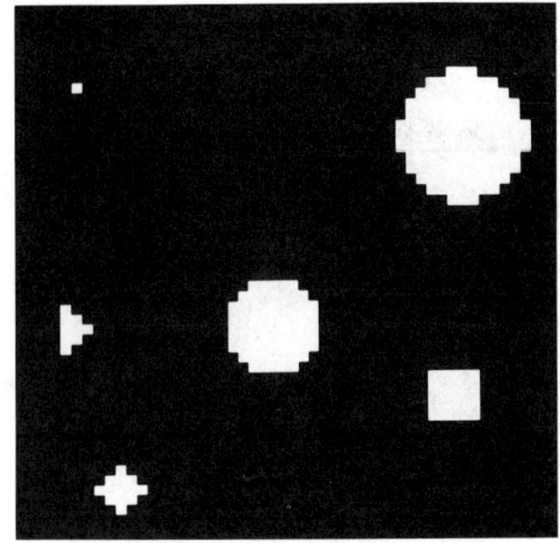

FIGURE 2.6. Dilation and Erosion by Disk 3

Erosion is anti-extensive, while dilation is extensive. For structuring element series which are strictly increasing, (such as disks of increasing size), erosion produces a *monotonically decreasing* series as a result, while dilation produces a *monotonically increasing* one.

The proof that dilation and erosion are monotonic with respect to an increasing structuring element sequence is fairly straightforward. Given an increasing stelt sequence $\mathcal{S}$, there must exist a T for each S_{i+1} such that $T = (S_{i+1} \setminus S_i)$, where '\' is the set subtraction operation, and so $S_{i+1} = (S_i \cup T)$. Then:

$$(X \oplus S'_{i+1}) = X \oplus (S_i \cup T)' = (X \oplus S'_i) \cup (X \oplus T')$$

$$(X \oplus S'_i) \cup (X \oplus T') \supseteq (X \oplus S'_i)$$

Likewise:

$$(X \ominus S'_{i+1}) = X \ominus (S_i \cup T)' = (X \ominus S'_i) \cap (X \ominus T')$$

$$(X \ominus S'_i) \cap (X \ominus T') \subseteq (X \ominus S'_i)$$

These results follow from the fact that the union is extensive with respect to its first argument, while the intersection is anti-extensive in the same way. For dilation and erosion then, the mere fact of having an increasing structuring element sequence guarantees the monotonicity property.

This fact allows us to classify pixels in a binary image, or 'voxels' in a grey image, according to the element of the stelt series for which they *first* appear (dilation) or disappear (erosion). The meaning of this classification is, strictly speaking, related to where each particular stelt in the series can

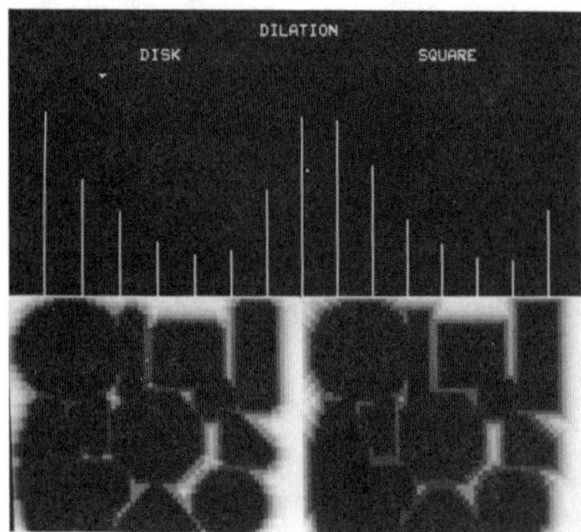

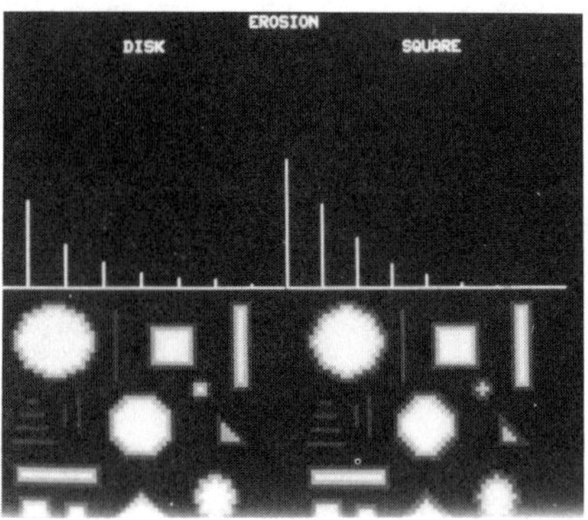

FIGURE 2.7. Coded Images and Distributions for Dilation & Erosion

be placed completely inside the foreground (erosion) or in the background (dilation). Erosion gives us as foreground all p such that S_p is contained in X, while dilation gives us as foreground all p such that S_p intersects X.

The meaning may become more specific, however, if we choose elements with special shapes. For example, if we choose a digital disk, the meaning becomes not just containment, but also distance—$X \circledwedge Disk_N$ is all of the locations at least distance N from the background, while $X \circledvee Disk_N$ is the set of points less than distance N from the foreground. If we use lines of a given orientation, we are now talking about distance along that orientation only. If we use vectors of a given direction, we are talking about distance in a given direction only. If we use squares or octagons, we can still think of it as a kind of distance, but of a different type—squares give us 8-connected distance, diamonds give us 4-connected, or city block distance, and octagons give us a mix between the two. Likewise, two point elements are the ends of lines or vectors, so they tell us about events at exact spacings and orientations, rather than within a certain distance at that orientation. *Although the basic operation has a meaning in terms of containment in the foreground or background, this meaning becomes specialized in character as we choose different structuring elements.* By knowing the specialized meaning, we can relate observations about image characteristics (criteria) to the kinds of operations which specifically address them.

Figure 2.5 shows an original binary image of some polygonal shapes, and Figure 2.6 shows its dilation and erosion by a disk or octagon of radius 3. The left hand side of Figure 2.7 shows coded images which represent the classification of each background pixel from the original image, based on dilation by disks and by squares. Also shown are the associated distributions for these coded images. A similar classification and pair of distributions for the foreground pixels of this image is illustrated on the right hand side for the erosion, also by disks and by squares.

Pseudocode for creating the coded images based on dilation and erosion is shown below.[6] Given the coded image, one simply takes a histogram of the non-zero pixels to get the associated distribution. Pixels with value 1 are of class 0, value 2 of class 1, and value N+1 of class N or greater. One can alternatively compute the distributions directly by simply counting the number of pixels which are different between successive pairs of dilations or erosions. The difference between result 'i' and result '$i+1$' becomes the distribution entry for class 'i'.

```
X0 = Original binary image (0/1 valued)

Dilation:                           Erosion:

Y = N+1                             Y = 0
FOR i = 0 TO N                      FOR i = 0 TO N
  X1 = X0 Dilate S(i)                 X1 = X0 Erode S(i)
  Y  = Y - X1                         Y  = Y + X1
ENDFOR                              ENDFOR
Return Y                            Return Y
```

Erosion and dilation also allow commutation and association of stelts used as sub-operators when they are joined by union or Minkadd (see Section 2.8). They distribute the union as defined in Equations 2.12 and 2.13. These facts allow us to split up certain erosions and dilations into equivalent sequences with fewer steps, as discussed in Section 2.8, or simply to rearrange the order of these steps. These properties are not true for all morphological operations, as we will see in the next section.

The meaning of the erosion and dilation for grey images is roughly the same but their effect is quite different than in the binary image case. One way to think of it is to imagine the grey level surface as a landscape, and then to pass the structuring element below the surface (the foreground), or over it (in the background). We can think of each pixel in the 2D support as having a number of little cubes (voxels) stacked on top of it, representing its height or grey value. These cubes will remain after an erosion if they represent locations where the stelt could be centered such that it would fit entirely below the surface (see Figure 2.8). For a dilation, the cube remains

[6]Note that while correct and conceptually simple, these are not necessarily the fastest algorithms for computing these images.

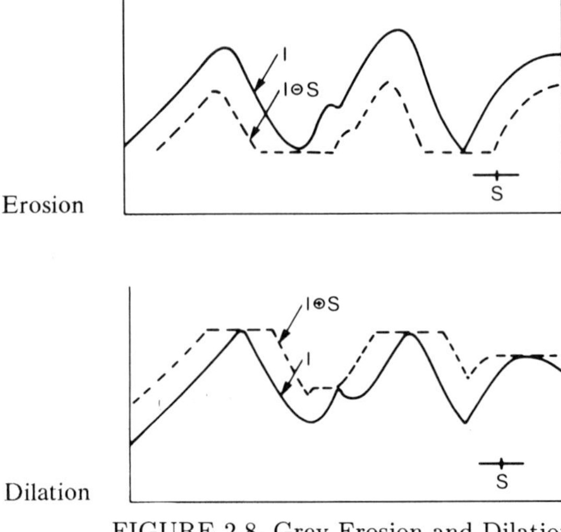

FIGURE 2.8. Grey Erosion and Dilation

at the end if it does *not* represent a location where the stelt could be placed to fit entirely above the surface (in the background). The interesting thing, which makes this more complicated than the binary case, is that while the structuring elements work primarily according to criteria related to containment in the X and Y dimensions, their effect is to modify the Z level or grey dimension, almost as a side effect.

We can alternatively think of the grey value of the erosion at any pixel as the *maximum* value for which the structuring element centered at that point and level, still fits entirely within the foreground under the surface. This is computed by taking the *minimum* of the grey surface translated by all the points of the structuring element. The dilation value is the maximum value for which the stelt (turned upside-down) at that location does *not* fit entirely within the background above the surface. This is computed as a *maximum* of translations of the grey surface. These definitions are true both for 'flat' stelts and ones which have grey level offsets at each point. Another way of looking at it for flat stelts is that the erosion returns the minimum value within the region covered by the structuring element, while the dilation returns the maximum. Grey stelts modify this by biasing the neighbors by different constants before taking the minimum or maximum.

The point about turning the element upside down for the dilation is a little tricky, especially since the definitions seem to indicate the opposite (reflection in Z is performed in the erosion, not the dilation). The idea is that to find out where a positive valued element fits under the surface we have to translate the image in the opposite direction (i. e., to reflect the stelt), and take the Min, while to find out where such an element cannot fit in the inverted sense above the surface requires reflecting once to get the inverted stelt, and once more because we once again have to translate in the opposite direction, this time taking the Max. In this latter case the two reflections cancel which is why the dilation definition does not include a reflection.

The variation of the meaning of the erosion and dilation operations as a function of structuring element shape as described above is roughly the same for the grey case too—the basic notion is one of containment, but becomes specialized with disks telling us about isotropic distance, lines about linear distance in a given orientation, vectors about distance in a given direction, and so on. The 'distance' here is usually distance to the grey level surface along a horizontal (Z = constant) plane. If we use true grey level structuring elements such as balls or cones or ellipsoids, the notions of distance and containment extend to more than just the horizontal plane.

This summarizes the definition, properties, and meanings of the most common erosion and dilation operations, which form the basis for almost all other morphological operations. In the following section, we will look at what happens when we combine erosions and dilations together in sequence.

2.11 Openings and Closings

The most important pair of morphological operations after the erosions and dilations are the openings and closings. Speaking generally, the opening is an erosion followed by a dilation, while a closing is a dilation followed by an erosion. Formally, the definitions are as follows:

$$(X \circledcirc S) = ((X \mathbin{\bigcirc\!\!\!\!\wedge} S) \mathbin{\bigcirc\!\!\!\!\vee} S') = ((X \ominus S'^{\wedge}) \oplus S)$$

$$(X \odot S) = ((X \mathbin{\bigcirc\!\!\!\!\vee} S) \mathbin{\bigcirc\!\!\!\!\wedge} S') = ((X \oplus S') \ominus S^{\wedge})$$

The important point to remember is that the dilation inside the opening is really a Minkowski addition by the same element used for the erosion, while the erosion inside the closing is really a Minkowski subtraction by the same element used for the original dilation, (except that grey level stelts require that we add a reflection). If we had not used the Minkowski operations at the end, the result would be operations which do not have the special properties that the openings and closings have, except with symmetric structuring elements, in which case they are equivalent to the latter.

The opening and closing are dual operations which are also increasing. Opening is anti-extensive, while closing is extensive. The most important new property of the opening and closing is that they are idempotent, that is:

$$((X \circledcirc S) \circledcirc S) = (X \circledcirc S)$$

$$((X \odot S) \odot S) = (X \odot S)$$

The proof for the opening is quite simple:

$$(X \circledcirc S) \supseteq ((X \circledcirc S) \circledcirc S) \quad \text{(Anti-extensivity)}$$

$$((X \circledcirc S) \circledcirc S) \supseteq (X \circledcirc S) \quad \text{since}$$

$$((X \circledcirc S) \circledcirc S) = ((((X \ominus S) \oplus S) \ominus S) \oplus S) = (((X \ominus S) \odot S') \oplus S)$$

$$\supseteq ((X \ominus S) \oplus S) = (X \circledcirc S)$$

The containment follows from the fact that closing is extensive and the Minkowski addition is increasing.

This means that for these two operations, additional applications of the same structuring element will not change the image. Several important consequences of this follow, first that if S_2 is *similar* up to a dilation to S_1, meaning that there is a T, such that $S_2 = (S_1 \oplus T)$,[7] then:

$$((X \circledcirc S_1) \circledcirc S_2) = (X \circledcirc S_2) = ((X \circledcirc S_2) \circledcirc S_1)$$

$$((X \odot S_1) \odot S_2) = (X \odot S_2) = ((X \odot S_2) \odot S_1)$$

The proof of the first equality for the opening:

$$(X \circledcirc S_2) \supseteq ((X \circledcirc S_1) \circledcirc S_2) \quad \text{(Anti-ext., Increasing)}$$

$$\text{and} \quad ((X \circledcirc S_1) \circledcirc S_2) \supseteq (X \circledcirc S_2) \quad \text{since}$$

$$(X \circledcirc S_2) = ((X \circledcirc S_2) \circledcirc S_2) \quad \text{(Idempotence)}$$

$$= (((X \ominus (S_1 \oplus T)) \oplus (S_1 \oplus T)) \circledcirc S_2)$$

$$= (((((X \ominus S_1) \ominus T) \oplus T) \oplus S_1) \circledcirc S_2)$$

$$= ((((X \ominus S_1) \circledcirc T) \oplus S_1) \circledcirc S_2)$$

$$\subseteq (((X \ominus S_1) \oplus S_1) \circledcirc S_2) = ((X \circledcirc S_1) \circledcirc S_2)$$

The final containment follows by the fact that opening is anti-extensive, and because both Minkowski addition and the opening operations are increasing. In effect we are showing that $(X \circledcirc S_2) \subseteq (X \circledcirc S_1)$, due to the fact that S_2 is open with respect to S_1, or similar to it up to a dilation. The proofs for $((X \circledcirc S_2) \circledcirc S_1)$ and for the closing identities are similar.

What this means is that the order of the openings or closings makes no difference; we can commute or associate them as we wish, but this is wasted effort since it is always the stronger filter (based on the larger stelt)

---

[7] An equivalent condition which Serra refers to more often is that S_2 be *open with respect to* S_1, meaning that $S_2 = (S_2 \circledcirc S_1)$.

which determines the result, assuming that the similarity condition holds. In this sense the opening and closing operations act like sieves—if I use a fine sieve and then a coarse one on the result, it is the same than if I reverse the order, or if I just use the coarse one—the result is what cannot pass through the coarse sieve.

If the similarity condition does not hold, then the opening and closing are in general neither commutative nor associative. In addition, we cannot distribute structuring elements with opening and closing as we did for erosion and dilation. $(X \circledcirc (S_1 \cup S_2))$ cannot be broken into the union of two openings, and $(X \circledcirc (S_1 \bigoplus S_2))$ cannot be expressed as a cascaded sequence of openings by the smaller elements. As a result, the only way we can speed up openings and closings is by decomposition of the erosions and dilations from which they are constructed.

In essence, what the openings and closings tell us about an image are the *sizes* of things—the sizes of objects in the foreground, or spaces in the background. They let us classify pixels according to the size (and/or shape) of the region to which they belong, rather than according to their distance from the foreground or background, as was the case with the erosion and dilation. Expressed formally, the binary opening gives us:

$$\{p \mid p \in (S_t \subseteq X), \forall_t\}$$

while the binary closing gives us:

$$\{p \mid p \notin (S_t \subseteq X^c), \forall_t\}$$

In English, the opening tells us where whole shapes fit (not just where they can be centered), while the closing tells us where they fit in the background. This gives us a direct method for addressing certain shape criteria. The opening will eliminate small 'islands' in a binary image, and suppress narrow 'isthmuses' and 'capes'. The closing will fill in small 'lakes' and suppress narrow 'channels' and 'bays'. Figure 2.9 illustrates the closing and opening of the binary image shown earlier by a disk or octagon of size 3.

The notion of size comes about if we think in terms of a series of the same shapes with different sizes such as a series of disks, squares, or lines—the opening and closing tell which sizes of these shapes will fit in the foreground or background and where. As with the erosion and dilation, the exact meaning of a given operation depends on the structuring element used. Disks will give us an isotropic measure of size, or can allow us to look for disk-shaped objects. Octagons and squares can be used to give a faster, rough indication of size, or to search for objects with these shapes. Lines tell us about size in certain orientations. Point pairs can be used to select periodic structures of a certain size and orientation.

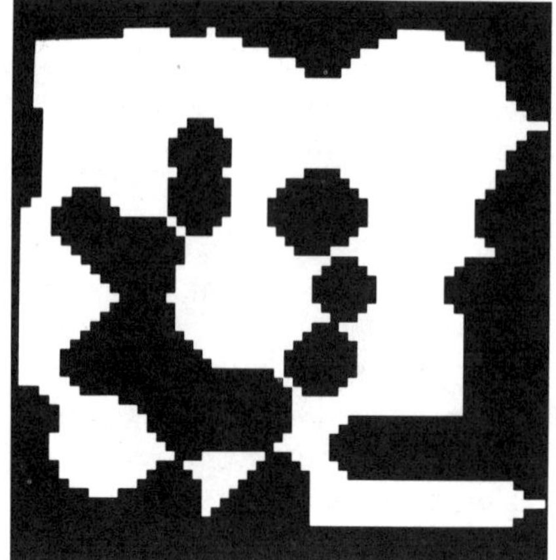
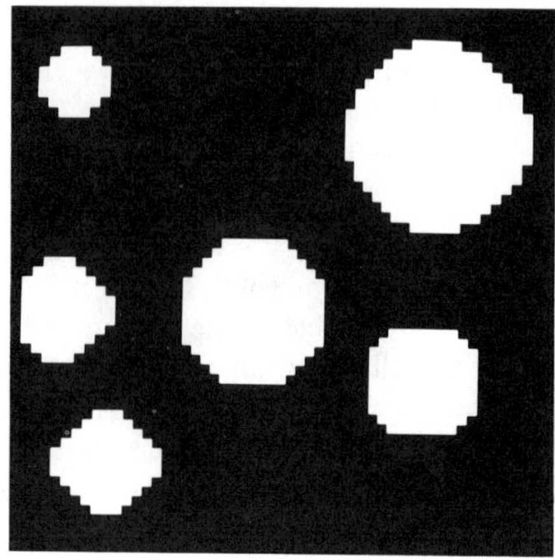

FIGURE 2.9. Closing and Opening of Binary Image by Disk 3

One other property of openings and closings which differs from the erosions and dilations is that the former are 'origin independent', that is:

$$(X \circledcirc S) = (X \circledcirc S_t)$$

$$(X \odot S) = (X \odot S_t)$$

This follows from the fact the opening consists of both an erosion and a Minkowski addition. While a translation of the structuring element in either of these operations produces a translated result, the translations are in opposite directions and cancel each other when we combine them, i. e.:

$$((X \ominus S_t) \oplus S_t) = [(([X \ominus S]_{t'}) \oplus S]_t$$

$$= [[(X \ominus S) \oplus S]_{t'}]_t = (X \ominus S) \oplus S$$

since $[t']_t = \mathcal{O}$.

The meaning of the opening and closing for grey level images is not too different than for binary ones; as with the erosion and dilation, these operations commute with thresholding if we use flat structuring elements. Essentially, the result for the opening for all cases will give us all the voxels which belong to instances of the structuring element which fit in the foreground, and the closing gives all of them which do not belong to instances which fit in the background. For flat elements this corresponds to taking the binary opening or closing at each individual grey level, and stacking the results on top of one another.

In more visual terms the opening cuts down the peaks in the grey level topography to the highest level for which the element fits under the surface.

The closing fills in valleys to the maximum level for which the element fails to fit above the surface. This becomes somewhat more interesting when a true grey level stelt such as a half-sphere is used, for this tells us where this object 'rolls' under or over the topographic surface. Openings and closings can be used by themselves to suppress or filter grey level structures (peaks or valleys) which are undesired and small. Alternatively, by using these operations in conjuction with the set complement, as described below, they can be used to extract such structures, independent of larger variations in the grey level surface.

As with the erosion and dilation, if we take an increasing series of structuring elements and apply each one to an image the opening will (in general) give a monotonically decreasing series as a result, while the closing will (in general) give an increasing series.[8] Again this fact makes it possible for us to classify individual pixels in a binary image, or voxels in a grey level image, by when they first appear or disappear in such a series. (For example, we could use a series of squares of increasing size as stelts, and classify the pixels according to the size of the square for which they first disappeared after an opening).

For opening and closing, the form of proof for monotonicity given for the erosions and dilations does not work—if we replace dilation by closing or erosion by opening in the previous equations, and expand the terms, we eventually arrive at a complicated expression which cannot be further reduced, and which cannot be proven to satisfy the necessary containments. While this might seem surprising at first, it is not too hard to understand why this is true if we look at an extreme case, as illustrated in Figure 2.10.

Here we see a disk shape inscribed within a square. Suppose that these are two structuring elements in our increasing sequence $\mathcal{S}$. Certainly the square contains the disk, but what is going to happen when we try to open by one or the other? If we try to open a large square or rectangle by the disk, it will not fit into the corners so these will be eliminated. If we open by the square, however, it will fit everywhere so that nothing is lost, despite the fact that it contains the disk! As this simple example illustrates, containment alone is not enough to guarantee monotonicity for openings and closings. We have to use a stronger condition.

As it turns out, the required condition to obtain monotonicity for openings and closings is again that the elements of sequence $\mathcal{S}$ be *similar* to one another, up to a dilation,[9] or in other words:

$$\forall_i \exists_T \mid S_{i+1} = (S_i \oplus T)$$

---

[8] One major exception to this rule occurs with disk-shaped structuring elements, discussed next.

[9] Or *open* with respect to one another.

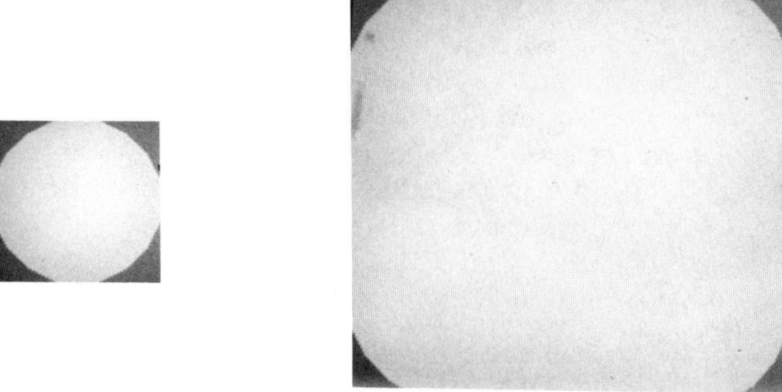

FIGURE 2.10. Disk Inscribed in Square and Openings of a Large Square

Given this, the proof of monotonicity for closing is as follows:

$$(X \odot S_{i+1}) = X \odot (S_i \oplus T)$$

$$= (X \oplus (S_i \oplus T)') \ominus (S_i \oplus T)$$

$$= (X \oplus (S_i \oplus T)') \ominus (T \oplus S_i)$$

$$= (((X \oplus S_i') \oplus T') \ominus T) \ominus S_i$$

$$= ((X \oplus S_i') \odot T) \ominus S_i$$

$$\supseteq (X \oplus S_i') \ominus S_i = (X \odot S_i)$$

The containment comes from the fact that closing is extensive and Minkowski subtraction is increasing. Specifically when we close $(X \oplus S'_i)$ by T, we get a result which contains the original since closing is extensive. But if this is true, then when we apply $\ominus S_i$ to both sides the containment remains true, because Minkowski subtraction is an increasing operation.

A similar argument holds for the opening, as follows:

$$(X \circledcirc S_{i+1}) = X \circledcirc (S_i \oplus T)$$

$$= (X \ominus (S_i \oplus T)') \oplus (S_i \oplus T)$$

$$= (X \ominus (S_i \oplus T)') \oplus (T \oplus S_i)$$

$$= (((X \ominus S_i') \ominus T') \oplus T) \oplus S_i$$

$$= ((X \ominus S_i') \circledcirc T) \oplus S_i$$

$$\subseteq (X \ominus S_i') \oplus S_i = (X \circledcirc S_i)$$

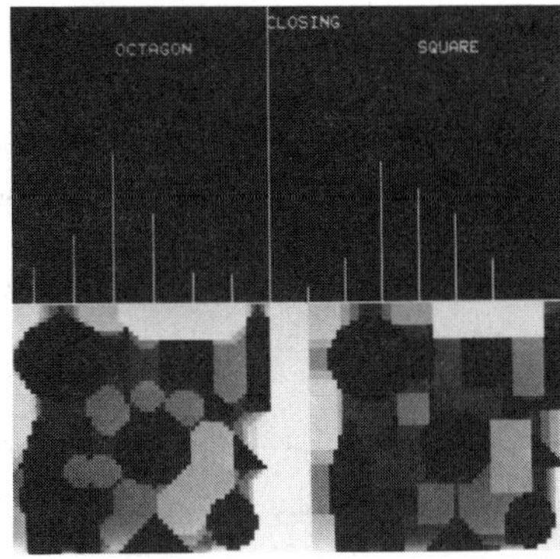

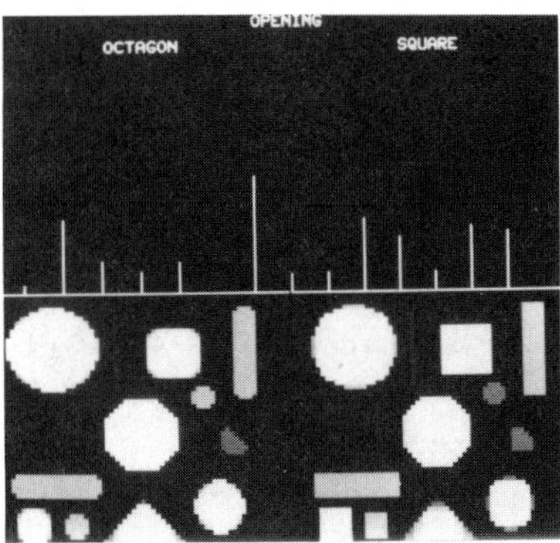

FIGURE 2.11. Coded Images, Distributions for Closings and Openings

The final containment depends on the anti-extensivity of the opening and the increasing property of the Minkowski addition, which preserves containment when we apply the same S_i to both sides.

Figure 2.11 illustrates coded images and distributions for closings and openings of the original image using octagons and squares, both of which satisfy the condition of 'similarity up to a dilation'. In fact, lines, vectors, and diamonds on a square grid, and hexagons and dodecagons on a hexagonal grid, also satisfy this condition and can thus be used to generate opening and closing distributions without any problem. The pseudocode for generating the coded images in these cases is listed below—the algorithms are identical to those for the dilation and erosion. All of these algorithms assume the monotonicity properties (inclusion of successive results) in order to work properly.

```
X0 = Original binary image (0/1 valued)

Closing:

Y = N+1
FOR i = 0 TO N
  X1 = X0 Close S(i)
  Y  = Y - X1
ENDFOR
Return Y
```

```
Opening:

Y = 0
FOR i = 0 TO N
  X1 = X0 Open S(i)
  Y  = Y + X1
ENDFOR
Return Y
```

The problem with a sequence of digital disks is that, unlike digital octagons, squares, diamonds, lines, and even Euclidean disks, such a sequence does not satisfy the condition of similarity up to a dilation for all successive pairs. The problem first manifests itself with disks of size 3 and 4 (see Figure 2.1). Though disk 3 and disk 4 as illustrated are correct, in that they represent the closest digital approximations to Euclidean disks of the same radius, it is not possible to obtain disk 4 by a dilation of disk 3—specifically, it is not possible to obtain the *surface shape* of disk 4 by starting with the disk 3 surface.

When we look at our original image (Figure 2.5), we can find a few locations (upper left and lower right figures) where the shape of disk 4 more closely matches the foreground edge shape, than disk 3. In these locations the opening by disk 4 will actually recover two more pixels than the opening by disk 3, and that will destroy the expected inclusion relationship, which is that the opening by disk 4 should be contained in the opening by disk 3. What this means in terms of the distributions is that some pixels have ambiguous classifications because they fall into more than one category. The pixels mentioned above disappear when we go from disk 2 to disk 3, they re-appear with disk 4, and then disappear again when we go from disk 4 to disk 5. Should they be classified as size 2 or size 4? Moreover, if we use the usual algorithm to generate the coded image we will get errors at these points when we try to 'stack' individual results on one another by adding or subtracting, since the inclusion relation does not hold. How could we modify the algorithm to eliminate such errors?

It is not enough to simply ignore this problem since correct digital disks are the most useful structuring elements in terms of accuracy and rotation independence. In many imaging problems, either the orientation of objects cannot be controlled, or else the objects themselves have shapes or widths which may vary with respect to direction, so one needs the ability to make size measurements which are isotropic. In some cases the use of octagons (or better, dodecagons in hexagonal space) provide a sufficient approximation to digital disks and avoids the difficulty described here, but in others the use of disks is absolutely required.

As it turns out, the number of pixels which have ambiguous classifications as a result of using digital disks in a given image is generally small and they tend to occur in dispersed locations. In contrast to the example of the square and inscribed disk described above, digital disks in a sequence which differ by only one value in radius are very nearly the same shape. As a result, the ambiguous pixels never occur in large clusters but are instead isolated to one or two pixels in a given location. They also tend to be rare events, since the similarity requirement does hold for many of the pairs of disks in the digital sequence. For these reasons, one can justify an approach to dealing with the problem posed by digital disks, which simply suppresses all secondary classifications of a pixel, once it has been initially classified.

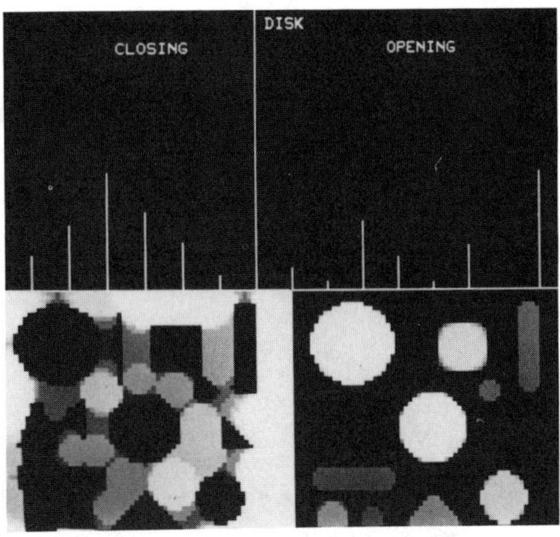

FIGURE 2.12. Coded Images, Distributions for Disk Closings & Openings

This can be done as shown below by making a very small modification to the code presented previously, which in effect checks to see if a pixel has already been classified, and if so prevents it from being re-classified. The result of applying these algorithms for disk-based closings and openings, to our original image, is illustrated in Figure 2.12.

```
X0 = Original binary image (0/1 valued)

Closing(revised):

Y = N+1
FOR i = 0 TO N
  X2 = 1 if (Y <= N)
         else 0
  X1 = (X0 Close S(i))
       Union X2
  Y  = Y - X1
ENDFOR
Return Y
```

```
Opening(revised):

Y = 0
FOR i = 0 TO N
  X2 = 1 if (Y >= i)
         else 0
  X1 = (X0 Open S(i))
       Intersect X2
  Y  = Y + X1
ENDFOR
Return Y
```

2.12 Residues

Now that we have defined the four most important morphological operations of erosion, dilation, opening and closing, we want to examine a new class of operations based on the use of these with the union and intersection complement operations. Called 'residues', these operations give us the difference between an original set and the set produced by one of the four morphological operators, as follows:

$$\begin{aligned}
(\text{X Eroderes S}) &= X \cap (X \owedge S)^c = X \cap (X \ominus S')^c \\
(\text{X Dilateres S}) &= X \cup (X \ovee S)^c = X \cup (X \oplus S')^c \\
(\text{X Openres S}) &= X \cap (X \circledcirc S)^c = X \cap ((X \owedge S) \oplus S)^c \\
(\text{X Closeres S}) &= X \cup (X \odot S)^c = X \cup ((X \ovee S) \ominus S)^c
\end{aligned}$$

In each case, the residue operation gives us the difference between the original image and the result of its transform.[10] For binary images, this difference is just the pixel locations which were added or deleted from the original by the transform. The importance of meaning or these operations is that they express the *complement* of the meaning expressed by their base operations. For example, the erosion residue gives us all pixels in X which were *not* locations where S could be centered such that it was contained entirely in X:

$$\{p : p \in X \wedge S_p \not\subseteq X\}$$

If S is a disk then this translates to all pixels which are *less* than a certain radius from the background; if a square then those which are less than a certain number of pixels away; if a diamond then those which are less than a certain city block distance; if a line then those which are less than a certain distance in the given orientation; if a vector then those less than a certain distance in the given direction, if two points then those without a neighbor at a given distance and orientation, and so on. The erosion by itself gave us the same type of criterion in each case but expressed in terms of 'greater than or equal' rather than 'less than'.

Similarly the opening residue for a disk of radius R gives us all pixels which belong to a region whose isotropic size is less than R, while the opening itself gave all those for which this size was greater than or equal to R. Other opening residues give us pixels whose size category is less than some value, for other kinds of shapes. The closing residues, which are defined completely symmetrically, address analogous size criteria for pixels

[10]The dilation and closing residues defined here differ from their usual definitions found elsewhere—this was done to make them symmetric and dual to the erosion and opening residues. See Section 6.1 for further discussion of this point.

in the background of the binary image. The opening residue lets us extract the small 'islands', narrow 'capes' and 'isthmuses' which were too thin to pass the size/shape criterion enforced by the opening, while the closing residue extracts the 'lakes', 'bays', and 'channels' of the background by the same type of criterion, for binary images.

For grey level images the interpretation is slightly different, because the differencing operations for grey images are based on the 'clipped addition' and 'clipped subtraction', defined in Section 2.5. The definitions of the four grey level residue operations are thus as follows:

$$
\begin{aligned}
(\text{X Eroderes S}) &= X \mathbin{\cup\!\!\!\!\!-} (X \owedge S) \\
(\text{X Dilateres S}) &= X \mathbin{\cap\!\!\!\!\!+} (X \ovee S)^c \\
(\text{X Openres S}) &= X \mathbin{\cup\!\!\!\!\!-} (X \circledcirc S) \\
(\text{X Closeres S}) &= X \mathbin{\cap\!\!\!\!\!+} (X \odot S)^c
\end{aligned}
$$

Turning to specific cases, the erosion residue computes the difference between each pixel value and the value of the local minimum within the structuring element region, or in other words the maximum drop from that pixel to any of the points within the region. The dilation residue computes the difference between each pixel value and the value of the local maximum within the region, or in other words the maximum climb from that pixel within the region. These residues can also be used to compute gradients, contours, and borders.

The residues of the openings and closings give us the voxels for which the structuring element failed to fit up in the foreground, or down in the background. In general, the opening residues give us peaks and ridges in the grey topography which were too small or narrow to contain the stelt, while the closing gives us sinks and valleys in the background which were too small to contain it from above. These operations are quite important for dealing with grey level images since usually the things we want to find appear as locally bright or dark regions in the image. The opening and closing residues allow us to 'normalize' the background and extract such regions, based on their size characteristics, independent of the absolute grey level values which compose them.

In terms of properties, the major significant ones which hold are that the erosion and opening residues are monotonic increasing, while the dilation and closing residues are monotonic decreasing.[11] They not increasing, and not preserving of homotopy. They don't commute, don't associate,

[11] Only under the right conditions for the opening and closing, as discussed in 2.11.

and don't distribute. The opening and erosion residues are, however, anti-extensive while the closing and dilation ones are extensive. The opening and closing residues are also idempotent.

While they don't preserve some of the nice properties of their base operations, the residues are nevertheless extremely important for their ability to extract certain kinds of grey level variations, and their use in classifying pixels, including their ability to express the complement of the conditions expressed by their base operations.

2.13 Band Operations

In much the same way that a threshold operation separates grey pixels which are greater than or equal to a certain level from those which are less, the erosions, dilations, openings and closings perform such discriminations based on spatial proximity criteria. The base operations are similar to upper thresholds (greater than or equal to) while the residue operations are analogous to lower thresholds (less than). Just as it is possible to create 'band' threshold operations which select pixels with grey values within a certain range, so it is possible by the use of basic morphological operations and their residues to create 'band-pass' morphological operators for selecting pixels based on shape criteria, by making use of the fact that they satisfy the monotonic properties.[12]

The meaning of the band pass operation is the conjunction of the separate parts, for example, if S_1 is a disk of size 5 and S_2 is one of size 10, then the band pass erosion operator gives us all pixels which are greater than or equal to distance 5 from the object edges and less than distance 10 (on a binary image). We can define general band operations which take a stelt shape and two sizes as arguments, such as (X Band-Erode Shape N1 N2) with the convention that if N1 is NIL or 0, then we compute the residue operation using N2, and if N2 is NIL or 0, then we compute the base operation using N1. This gives all of the combinations of the base operation and its residue in a single operation, much like the the idea of the band threshold operation with respect to grey levels.

The meanings of the band operators for binary erosion, dilation, opening and closing are fairly straightforward once their base operators are understood, including the influence of the structuring element shape being used on the meaning. I will not explain them further; see Sections 2.10 and 2.11 for descriptions of those operators and their meanings. Because of the difference in the interpretation of the set combination operations for grey

[12]See Section 4.1 for further details and Appendix B for formal definitions of several different classes of band operators.

images, however, the meanings of the grey band operations deserve some further elaboration.[13]

The meaning of the grey level band operators for each pixel is really the difference in grey levels between the value for which it satisfies one condition and the value for which it satisfies another. For example, the band operator using erosion with a disk structuring element between values 5 and 10 gives us the difference in grey values between the level at which the pixel was at least distance 5 from the grey level surface (laterally within the foreground) and the level at which it was at least distance 10 away. (Perhaps the pixels we want have a large difference between these two levels while those we don't want have a smaller one).

For the opening using the same parameters, the meaning would be the difference between the level for which the pixel belonged to a horizontal section of the foreground of size 5 or greater, and the level for which it belonged to one of size 10 or greater. To get the meanings for the dilation and the closing, we simply change foreground to background using the same basic criteria.

It may seem that the meanings of these grey level band pass operators are a bit esoteric and difficult to apply. That is true. They are not very often used for the reason that it is difficult to see from just looking at an image, when they would be of particular use. Part of the problem comes from the fact that the criteria they use are based on spatial properties in the X-Y plane, but their effect is to select subsets of voxels along the Z dimension, different ones for different X-Y pixels. These are then stacked up together on the zero plane so that all of the previous voxel adjacency relations are destroyed. Even though they may seem hard to use, however, the grey band operators do provide us with additional tools for discriminating pixels, which may be actually easier for computers to apply than the human visual system.

The fact that we can generalize the normal base operations and their associated residue operations to specific instances of band operations, means that we can homogenize and simplify the process of evaluating each alternative by thinking only in terms of band operations in our search process. That is, instead of thinking about openings and opening residues separately, we will only think in terms of finding the best band-opening, which will in some cases reduce to one of the other two.

---

[13] See Appendix B for complete definitions.

2.14 Particle and Hole Operations

Now that we have defined the basic four morphology operations, along with their residues and related band operations, I want to look at three other pairs of operators which work on whole connected components (particles in the foreground or holes in the background). In a sense these operators classify individual pixels according to properties of the entire object to which they belong. All of these operations are based on conditional dilation (for particles) and conditional erosion (for holes):

$$\text{(M CDilate B)} = (M \overset{\oplus}{\rightarrow} B) = \forall_q \mid \exists_{P \in B}, q \in P \wedge (M \Uparrow P)$$

$$\text{(M CErode B)} = (M \overset{\ominus}{\rightarrow} B) = \forall_q \mid q \in B \vee \exists_{H \in B^c}, q \in H \wedge (M \supseteq H)$$

The conditional dilation and erosion operators are really set combination operators, but unlike the union and intersection, their arguments play different roles. The first argument is a 'marker' set which is used to 'tag' certain particles or holes in the second argument or 'base' set, in order to keep them. All others are eliminated. In effect, conditional dilation and erosion address the criterion of *connectedness*. Conditional dilation is anti-extensive with respect to the base set B, because it removes particles in B which were not hit by M. Likewise, the conditional erosion is extensive with respect to B, because it retains only the holes in B which are 'marked' by holes in M, and fills up all of the others.

If we replace M in the conditional dilation operator above, by an anti-extensive operator on B, then the result gives us a unary anti-extensive operation on B which selects certain particles of it based on the specific effect of the anti-extensive operator. Likewise, by replacing M in the conditional erosion operator by an extensive operator on B, we get a unary extensive operation on B which selects (retains) certain holes in B based on the effect of the chosen extensive operator. Then, by combining these new operations using different parameters with unions, intersections, and complements, we can define new extensive and anti-extensive band operations which can be used to choose *ranges* of particles or holes, based on a classification of them as entire connected components. I will not elaborate on these definitions here, but several different band operators of this type are defined in abstract form in Appendix B.

For the purposes of the implementation, I have chosen three types of band operations which work on particles or holes. The first pair address the issue of the *size* of the particle or hole, for example:

$$Y = (X \circledwedge S) \overset{\oplus}{\rightarrow} X$$

$$Y = (X \circledvee S) \overset{\ominus}{\rightarrow} X$$

The second addresses the issue of the length of the particle or hole, in the sense of the longest internal (possibly curved) diameter. This one makes use of operations based on the digital thinning, but which continues to prune away at end points until the object disappears, if enough steps are taken. I call this the 'Reduce' operator, $\triangleright$. Its dual operation thickens and eventually eliminates holes in the background—I call this the 'Expand' operator, $\triangleleft$. The associated connected component operations tell us about the overall internal length of a particle or hole and are defined in the simplest case as follows, with S representing not so much a structuring element as a connectivity (4 or 8) and number of steps for the operation used:

$$Y = (X \triangleright S) \overset{\oplus}{\rightarrow} X$$

$$Y = (X \triangleleft S) \overset{\ominus}{\rightarrow} X$$

The third pair, which are slightly more complicated, are based on the set of residues of openings and closings of the particles and holes, respectively:

$$Y = (((X \cap (X \circledcirc S)^c) \circledwedge \varepsilon) \overset{\oplus}{\rightarrow} X)$$

$$Y = (((X \cup (X \odot S)^c) \circledvee \varepsilon) \overset{\ominus}{\rightarrow} X)$$

Here ε is a digital epsilon (a 2x2 square) which is used to eliminate small residues of 1–3 points as not relevant. These operators select particles and holes based on the size of the residue that they have with respect to a given sized structuring element. Depending on the one used, this can tell us something about their roughness and/or about their shape—since if their shape matches the structuring element exactly, there will be no residue. The same operator can also be used to distinguish polygons with sharp angles from those with more obtuse ones.

2.15 Grey Level Operations

The subject of morphological grey level operations is a very complex one, which cannot be addressed fully here. For the sake of completeness, however, I wanted at least touch on the issues of how grey level operations are used and chosen, and on some of the complexities involved in evaluating them. To simplify the problem for this project, I have decided to consider only grey operations which include a thresholding operation (i. e., return a binary image) and which therefore really act as operators for discriminating sets of pixels based partially on grey level criteria. This would include grey level erosion followed by a threshold, grey level closing residue followed by a threshold, and so on.

In actual practice, the most commonly used grey operations are of this type—they are used to pre-process the image so that a threshold operation will work better than it would have on the original image. The problem is that evaluating the utility of grey-to-grey operators is much more difficult especially if one sequences several together. The eye cannot evaluate them as well because we can only distinguish about 32 levels (5 bits) of grey, and automatically evaluating them requires much more calculation than for pure binary operations. The grey-to-binary operators provide a compromise.

Besides complexity, another reason that grey level operators are not used so often is that they require more complex hardware and/or take more time to execute. If one has a choice between grey-scale opening by a flat disk and then thresholding, or thresholding and then a binary opening, the latter choice is often much faster (Max/Min versus Or/And logic) on most machine architectures, but the answer is exactly the same. (Thresholding commutes with erosion, dilation, opening and closing by *flat* structuring elements).

What the grey operators really provide is more flexibility. The real difference between the two different sequences described above is that while thresholding first is often faster, the grey level opening effectively computes the binary opening on all 256 levels at once (faster than if doing these one at a time) and then lets one choose the best one afterwards by an adaptive process, which thresholding first makes impossible. If you are not sure of the best level at which to threshold from the beginning, the grey level operation makes it possible to make this decision as a function of the transformed image.

Grey level operations with non-flat structuring elements also allow some additional tests that would not be possible otherwise. These can give much cleaner results in certain situations, particularly if one can know and control the shape of important parts of the grey level topography. The most commonly-used non-flat structuring elements are the half-sphere and the cone.

The goal of binary operations is to correctly specify a certain set in the image plane, usually by classifying all of the pixels according to a structuring element series, and then choosing the parameters or bounds which do the best job of separating them into the desired foreground and background. In the grey level domain, this goal becomes modified to one of trying to maximally separate the histograms of the desired foreground and background, so that a simple threshold operation can be used to actually put them into different sets. In this case, for the same series of stelts used before in the binary situation, instead of keeping the *counts* of desired foreground and background pixels associated with each series element, one has to keep their *histograms*. Instead of looking for the size parameters which best separate the counts, one now has to look for those which best separate the histograms—a much harder task.

With all of this in mind, I want to define four classes of grey-to-binary operations which make use of the basic four grey level morphology operators and their residues. All of these are band operators, but in some cases the bands are in terms of grey levels, while in others they are in terms of a structuring element size parameter. (I have avoided operations which allow both the levels and the sizes to vary, because that immediately would increase the complexity of the search for optimal parameters to N^2 instead of N).

The first group of complex thresholding operations are those which perform a base grey level operation (erosion, dilation, opening, closing) and then select a grey level range from the result. The definitions, for $T_1 \leq T_2$ are:

$$(X \owedge S_\lambda)|_{T_1}^{T_2}$$

$$(X \ovee S_\lambda)|_{T_1}^{T_2}$$

$$(X \circledcirc S_\lambda)|_{T_1}^{T_2}$$

$$(X \odot S_\lambda)|_{T_1}^{T_2}$$

For flat structuring elements S, these operations are equivalent to thresholding X at T_1 and T_2, computing the binary morphological operation on both, and then taking the set difference between the two results, respectively:

$$(X|_{T_1}^{+} \owedge S_\lambda) \cap (X|_{T_2}^{+} \owedge S_\lambda)^c$$

$$(X|_{T_1}^{+} \ovee S_\lambda) \cap (X|_{T_2}^{+} \ovee S_\lambda)^c$$

$$(X|_{T_1}^{+} \circledcirc S_\lambda) \cap (X|_{T_2}^{+} \circledcirc S_\lambda)^c$$

$$(X|_{T_1}^{+} \odot S_\lambda) \cap (X|_{T_2}^{+} \odot S_\lambda)^c$$

The idea behind these operations is that the morphological transform will modify the grey levels of the pixels to make the desired ones more easily distinguished on the basis of grey level than was the case in the original image. For example, if both the desired and some undesired pixels have bright values, but the undesired ones exist in smaller, more compact clusters than the desired ones, then an opening by the correct sized stelt could cut down the undesired ones to lower values, without affecting the desired ones very much. The end result would be to increase the separability of the histograms of the desired pixels from the undesired ones.

In the case of these four operations, assuming that the stelt shape is fixed, the problem is to find the single size λ which provides the best overall histogram separation (maximum grey level difference), and then choose the specific thresholding parameters based on this. As indicated in the previous section, the criteria by which grey level modifications will be made depend on the type of operation, in conjunction with the stelt shape. Roughly speaking these will be: 'interior distance' from the grey level surface for the erosion, 'exterior distance' for the dilation, 'size' within the foreground for the opening, and 'spacing' within the background for the closing. The specific interpretation will also vary as a function of the stelt shape, as described in earlier sections. As an example, the interpretation for the opening by a disk would be to select pixels which belong to a column of radius $\geq \lambda$ at level T_1, but not at level T_2.

The second group of operations is similar to the first, but is based on the residues of morphological operations rather than the base operations:

$$(X \mathbin{\cup\!\!\!\!-} (X \owedge S_\lambda))|_{T_1}^{T_2}$$

$$(X \mathbin{\cap\!\!\!\!-} (X \ovee S_\lambda)^c)|_{T_1}^{T_2}$$

$$(X \mathbin{\cup\!\!\!\!-} (X \circledcirc S_\lambda))|_{T_1}^{T_2}$$

$$(X \mathbin{\cap\!\!\!\!-} (X \odot S_\lambda)^c)|_{T_1}^{T_2}$$

In the case of these operations, the grey level differencing prevents the possibility of performing the thresholds first, and then doing binary operations, even in the case of flat structuring elements. Here again, though, the goal is to look for the fixed λ which gives us the best overall separability of histograms after the residue operation. We then choose T_1 and T_2 to optimize the separation of the desired from the undesired pixels. Because we are using residue operations instead of base ones, the meaning of the result includes the negation of the similar operation from the first group.

For example, the opening residue by a disk case selects the pixels which do *not* belong to a column of radius $\geq \lambda$, and which are $\geq T_1$ above the maximum level for which a disk of size λ fits, but $< T_2$ above this level. This operation could be used if desired pixels belonged to smaller clusters of bright pixels (smaller peaks) than the undesired ones.

The third and fourth groups of complex thresholding operations work a little differently; they fix the threshold level, but use two different sized structuring elements of the same shape. The goal is to look at size or distance distributions and to pick the absolute grey level which gives the best overall separability in these between the desired and undesired pixels. Once this level is chosen, specific size parameters λ_1 and λ_2 are chosen to optimize the separation of the two pixel sets, similar to what was done above for the other two groups.

The third group is defined as follows:

$$(X \owedge S_{\lambda_1})|_T^+ \cap ((X \owedge S_{\lambda_2})|_T^+)^c$$

$$(X \textcircled{$\vee$} S_{\lambda_2})|_T^+ \cup ((X \textcircled{$\vee$} S_{\lambda_1})|_T^+)^c$$
$$(X \circledcirc S_{\lambda_1})|_T^+ \cap ((X \circledcirc S_{\lambda_2})|_T^+)^c$$
$$(X \odot S_{\lambda_2})|_T^+ \cup ((X \odot S_{\lambda_1})|_T^+)^c$$

For flat structuring elements these are equivalent to thresholding first and then taking the difference of the binary morphology operations by the different sized elements:

$$(X|_T^+ \textcircled{$\wedge$} S_{\lambda_1}) \cap (X|_T^+ \textcircled{$\wedge$} S_{\lambda_2})^c$$
$$(X|_T^+ \textcircled{$\vee$} S_{\lambda_1}) \cup (X|_T^+ \textcircled{$\vee$} S_{\lambda_2})^c$$
$$(X|_T^+ \circledcirc S_{\lambda_1}) \cap (X|_T^+ \circledcirc S_{\lambda_2})^c$$
$$(X|_T^+ \odot S_{\lambda_1}) \cup (X|_T^+ \odot S_{\lambda_2})^c$$

Here we are looking for some level T, which, if we perform two morphology operations of the same type but by different sizes, the difference in the two results will us the pixels we are looking for. As an example, for the opening by two disks we would get back all pixels which belong to a column at least T high and $\geq \lambda_1$ in radius, but $< \lambda_2$ in radius. In other words, the opening by the λ_1 disk at level T does not eliminate the pixel, but the opening by the λ_2 disk does.

This third group of complex thresholding operations is clearly related to the first. While it would also be possible two define both T and λ to vary over a range (4 parameters total), this would make the search process work over a two dimensional space which would be much more complex and time consuming, though certainly more capable.

The last group of thresholding operations to be defined here are related to the second in that they also use residue operations, and to the third, in that the level is fixed but two structuring element sizes are used. They are defined as:

$$(X \uplus (X \textcircled{$\wedge$} S_{\lambda_2}))|_T^+ \cap ((X \uplus (X \textcircled{$\wedge$} S_{\lambda_1}))|_T^+)^c$$
$$(X \Cap (X \textcircled{$\vee$} S_{\lambda_2})^c)|_T^+ \cup ((X \Cap (X \textcircled{$\vee$} S_{\lambda_1})^c)|_T^+)^c$$
$$(X \uplus (X \circledcirc S_{\lambda_2}))|_T^+ \cap ((X \uplus (X \circledcirc S_{\lambda_1}))|_T^+)^c$$
$$(X \Cap (X \odot S_{\lambda_2})^c)|_T^+ \cup ((X \Cap (X \odot S_{\lambda_1})^c)|_T^+)^c$$

These operations try to find a fixed level T, such that the difference in residues of this level by two different sized structuring elements, gives us the desired pixels. For the opening residue by a disk, we get those pixels which are $\geq T$ high after the residue operation by a disk of size λ_2, but not after the opening residue by one of size λ_1. The search requires comparing the size or distance distributions of the residue operations at each level T, for maximum overall separability, and then choosing the specific λ parameters at this level which give the best separation.

The subject of grey level operations is a very complex and rich one, offering many more possibilities to the algorithm developer in terms of ways to discriminate pixels, but at a cost of having do to a much more complex analysis of the data. Without going into great detail here, I nevertheless wanted to point out some of the varieties of operations that can be defined, the additional power that they offer, what their meanings are in terms of the different ways in which they can discriminate pixels, and what the real goal is of using them, in terms of finding ways to separate the histograms of the pixels of interest from those pixels which are not of interest.

2.16 Summary of Problem Domain

Before concluding this chapter, I want to summarize what I think are the most important concepts developed here. One of the major purposes of this chapter was to define the morphological and related operations that will be the foundation of the implementation effort described in Chapters 4 and 5. These include the thresholding operation, the four position mask operations, the erosions, dilations, openings, and closings by various structuring elements, and the connected component, or particle and hole operations. We saw not only their definitions, but also their specific meanings in terms of the criteria which they address in an image—something which is not often readily available in the literature concerning them. Specifically, the threshold addresses absolute brightness, the position mask operations concern absolute position in the image, the erosion and dilation speak to different notions of distance in the foreground and background, and the openings and closings speak to notions of size and spacing. The connected component operations provide ways of selecting particles and holes on the basis of size, roughness or shape, and overall length as measured by the longest internal diameter.

What this will mean in the implementation is that if we take a set of problem images, and we see by looking at them that we can separate the pixels we want from the ones we don't want, on the basis of any *one* of the image criteria mentioned above, then our automatic system for generating recognition algorithms described in Chapter 4 will be able to discover this, and generate the correct discrimination or recognition algorithms. This will be true, *no matter what image set we choose*, so long as the given criterion discriminates the desired and undesired pixels consistently in all of them.

Another very important concept developed in this chapter was the concept of *band* operations. If we take the basic operators discussed above, and combine the results for them with the set operations of union, intersection, and complement, it becomes possible to classify pixels in an image based on the effect of the underlying 'base' operator, and from this we can also define more powerful operators which allow us to select pixels based on the ranges of classes into which they fall. As Section 4.1 will show, these band operators are the basis of the implemented system's algorithm generation activity.

In the course of defining the band operations for this use, two significant obstacles had to be overcome. One was that the residue definitions for all extensive operations had to be changed to make them dual to the anti-extensive residues. The second was that openings and closings by disks did not have the monotonic property needed to correctly define band operators based on them—an alternative algorithm had to be developed for these cases.

Finally, I want to end this section by pointing out that there are many other operations commonly used in image processing, both morphological and non-morphological, which have not been addressed here. These include point transformations, pattern matching operations, skeletons and other topologically related operations, general filtering, smoothing, and texture operations, as well as gradient, contour/edge finding, and basic linear transformations such as convolutions and correlations. I touched on some of the grey operators above, but much more could be said concerning these operators as well. Each of these addresses different criteria in an image than the ones I have chosen to work with here. Nevertheless, the same principles apply and with more time and effort, all of them could be brought under a similar framework. Certain ones, for example linear convolutions and morphological grey level operations, would require some new work in terms of how they classify pixels; others, such as the contours and skeletons, might require other new ways to evaluate how well they perform. In principle, though, all of them could be subjected to the same kind of analysis, toward the goal of deriving the mathematical knowledge needed to apply them automatically, in the process of generating algorithms. While I have chosen the basic morphology operators to work with here because of some interesting properties which they have and because I am most familiar with them, the analysis used here is not entirely specific to them—it has ramifications for the generation of image processing algorithms in general.

3

Theory of Automatic Set Recognition

In this chapter I want to describe, both formally and informally, the process of algorithm development for set recognition tasks, including what are the essential components of this process and what are some of some of the techniques and strategies that can be used to improve results. The treatment given here will take the form of an overview; for certain topics additional detail will be found in Chapters 4 and 5, which describe the implementation and its current capabilities.

The chapter begins by discussing some basic terminology for describing such problems, followed by the presentation of a couple of simple examples. I then present a formal description of these types of problems, and present some arguments to show why they are exceptionally hard. In the next sections I discuss the problem of representing partial and complete algorithm states and present an overall description of the algorithm development process as performed by humans and its major components. The chapter closes with a discussion of some of the knowledge and methods used to speed up the search in the current implementation of REM, along with arguments that the methods used are complete, in that they will not miss any solutions which fall within their pre-defined scope.

3.1 Basic Terminology

Before discussing example algorithm development problems and the methods used in this activity, I want to first introduce some terminology. In trying to solve a set recognition problem or subproblem, it is useful to employ the concept of a *base* image or set (X), an *accept* set (A), and a *reject* set (R). At the beginning of a problem the base image X is just the original image, which we will designate in what follows as X0. The accept set A0 is all of the pixels we want to get out as a result, while the reject set R0 contains all of the pixels we definitely do *not* want in the result. Obviously the accept and reject sets must be disjoint, but it does not have to be the case that together they completely cover the image support space. We may be only interested in distinguishing one region from another; there may be other areas where we don't care about the results—these we will call 'ignore regions'.

Since most problems consist of not one, but a group or class of example images for which an algorithm must be found, what we really have at the beginning is a set of triplets: $X0_{1 \rightarrow N}$, $A0_{1 \rightarrow N}$, and $R0_{1 \rightarrow N}$, where N is the number of sample images in the input set for which we are developing the algorithm. The accept and reject masks for each original base image X0 may be constructed by having a human 'truth' the images by drawing with a light pen or trackball, or they may be pre-computed in some way. What we want to find is a *single* algorithm which, when applied to any $X0_i$, produces a result with *all* of the pixels in the set $A0_i$, and *none* of the ones in $R0_i$.

As the solution of a problem proceeds, the definitions of the base image and the accept and reject masks change. For example, if we have a partial solution to the original problem based on an operator Ψ, we still are left with a subproblem to finish the overall solution. The base image (X), for the new problem will be just the result of applying Ψ to X0, but the images or masks representing the A and R sets may have to be adjusted to reflect the fact that we can now ignore certain parts of the image (those already distinguished or 'determined' by the Ψ operator), and just concentrate on discriminating those parts of the A and R sets which were not separable by that operator.

In short, then, while we will use $X0_i$, $A0_i$, and $R0_i$ to describe the original images and their accept and reject masks, respectively, when we shift from this to talk about a particular subproblem, we will refer to its description in terms of the images X_i, A_i, and R_i, or just X/A/R for short.[1] In general a subproblem is solved when either its A set or R set has been determined, over all example instances.

With these terms in mind then, we are ready to consider some examples, and the general process of solving set recognition problems.

3.2 Example Problems

In this section, I want to go through a couple of examples to illustrate what whole algorithms look like, to give examples of the X, A, and R sets, and to consider what the steps are in developing an algorithm. Figure 3.1 shows a range (3D) image of a jumbled bin of parts. Each grey level pixel represents in effect a distance or height. The problem is that we have a two-fingered robot which can pick up the parts, so our task is to find all of the locations where the fingers can fit down around somewhere on a part

[1]While there is really a set of N instances for each, we will often just refer to a whole group as a single unit (X, A, or R), considering each group to be one of a triplet of constraints which define the given subproblem.

FIGURE 3.1. Range Image of Parts in a Bin (Courtesy ERIM)

sufficiently to get a good grip. This criterion can be sub-divided into three conditions: First, the part location has to be narrow enough to fit between the two fingers; second, there has to be enough space around the part for the fingers to fit down around; and third, there has to be enough of the part sticking up for the fingers to get a good grip.

The solution of the three condition problem just described can be expressed as three equations, each of which embodies one of the conditions:

$$X_1 = X_0 \cap ((X_0 \uplus (X_0 \circledcirc S_1))|_1^+)$$

$$X_2 = MIN_{[\theta = 0 \rightarrow \pi]}((X_0 \circledvee S_2) \circledvee S_{3_\theta})$$

$$X_3 = (X_1 \uplus X_2)|_H^+$$

Here X_0 is the original image, S_1 is a structuring element representing the space inside the robot fingers, S_2 is one which represents the size of one finger, and S_{3_θ} is a rotation of a two point structuring element with separation equal to that of the fingers. 'H' is the minimum overlap parameter.

The first equation finds all of the part locations in X_0 which can fit inside the space between the fingers; the second finds the places where both fingers can fit farthest down around the parts from above; and the third specifies that the overlap of these two at a given location be enough to get a good grip. X_3 is a binary image of candidate X-Y locations that meet all three conditions. By intersecting this set with the original X_0, we could return an image with just the heights of these candidate locations for the robot to select from. Here is an example of a fairly powerful grey level algorithm which can be expressed concisely as a series of morphological operations.[2]

We were able in this case to generate a solution to the problem based entirely on its description and a knowledge of the robot hand. For most problems, however, it is not possible to generate a solution so easily, either because we don't know enough about the problem domain, especially the contexts and backgrounds of the images, or else because there may be noise which interferes with our ability to do a direct solution, and which forces us to do some experimental testing. In this particular instance, for example, if we did not have the kind of knowledge about the problem as outlined, we might have to do some experiments in which we first took a number of images like that of Figure 3.1, and then determined manually where were the locations such that the robot could pick up the part. On the basis of this we could develop an accept mask of locations that we would like an algorithm to return when faced with a given image of this type. (Some or all of the other points would go into the reject mask). Following this 'truthing' procedure, we could begin developing algorithms to select the right locations for the robot given any such range image of the parts. We would do this by trying to find criteria in the images to distinguish the accept set of pixels from the rest. In the present case, these criteria would involve a combination of relative grey level height, thickness, and spacing on both sides of an area of interest.

As a second, simpler example, suppose we have a set of images like Figure 3.2, composed of disks, triangles, and rectangles, and suppose the goal was to accept all disks, reject all rectangles, and ignore everything else. In this case the accept masks would contain only the disks, and the reject masks would contain only the rectangles, including the squares, as shown. Next suppose we had found a way to reject all objects whose length and width differed—perhaps by testing the shape of the most extreme erosion of each particle for a non-negligable length. Removing these objects would partially solve the problem, but we would be left with the subproblem of distinguishing disks from squares. The description of this subproblem would consist of base images where the non-square rectangles had been removed, an accept mask still made up of the disks, and a reject mask containing the squares to be eliminated as the goal of the new subproblem. These three sets are shown at the bottom of Figure 3.2.

While this example illustrates the concept of accept and reject sets and how they can be modified when a subproblem is generated, there is actually a better way to solve this entire problem which depends on the observation that all of the objects in the accept set are round, while all of the other objects have some corners. This problem can be solved by accepting objects that have shapes similar to disks (i. e., which are *open* with respect

[2]See Lougheed and Sampson [1988], for an alternative solution to this problem.

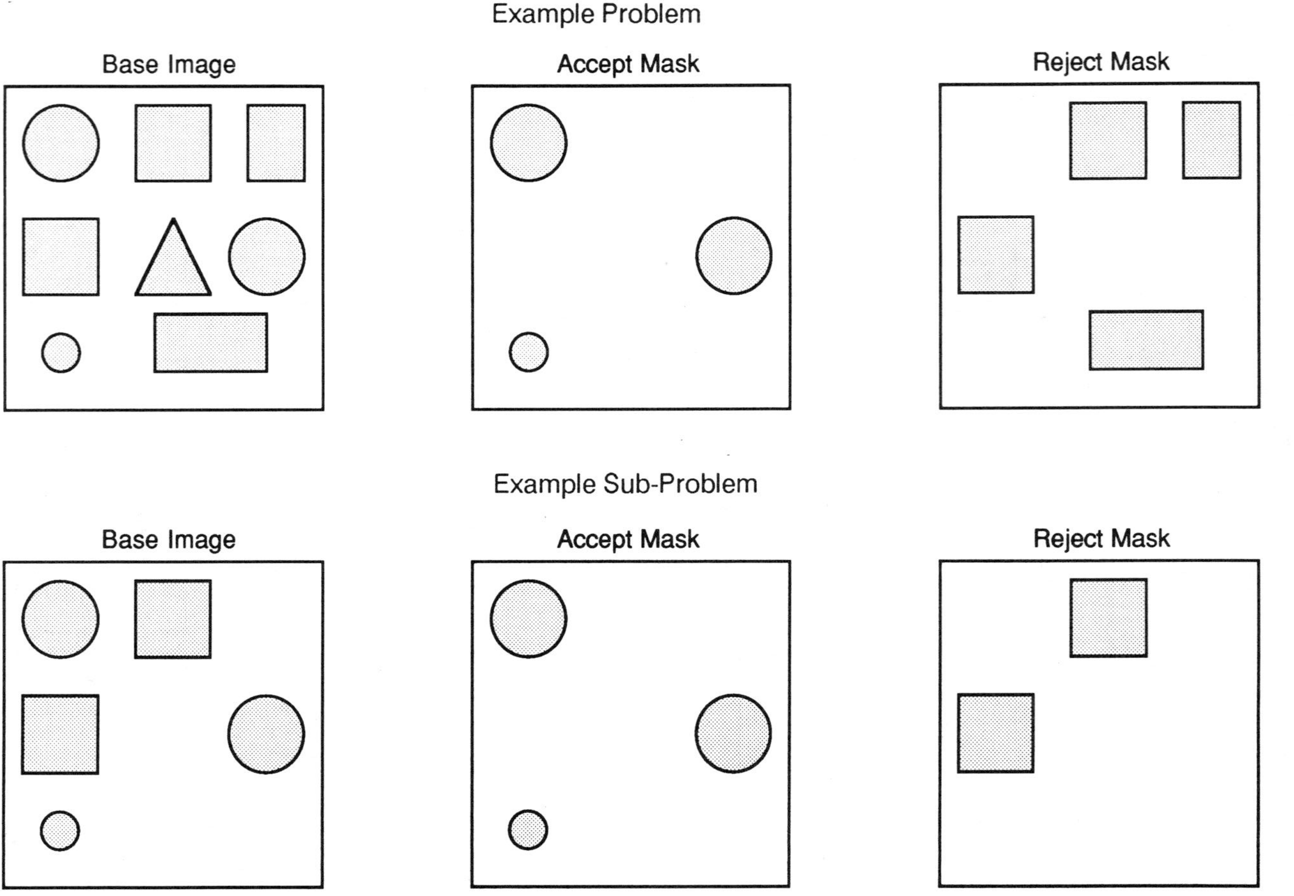

FIGURE 3.2. Binary Example Problem

to a small disk), and rejecting all others. An operator for doing this was described earlier in Section 2.14 based on the residue of the opening:

$$Y = (X0 \cap (((X0 \cap (X0 \circledcirc Disk_3)^c) \bigtriangleup \varepsilon) \overset{\oplus}{\rightarrow} X0)^c)$$

This operation first opens the image by a small disk and finds the residue. It erodes this residue by a very small (2x2) structuring element (ε), to eliminate small residues that might be due to digitization, rather than actual shape. Then it conditionally dilates this result over the original image to get all of the objects (non-disks) which had a significant residue with respect to the small disk. It then removes this set from the original to leave just the disks.

A question one might reasonably ask at this point is: How does one go about developing such an algorithm? In this case we already had a previously defined band operator defined for our purposes from Chapter 2, but suppose that wasn't the case? What is a likely sequence of steps that one would go through?

The answer to this question begins with the observation that all of the pixels we want are already part of the foreground. This tells us that we want to consider an anti-extensive operation, one which only *removes* pixels. Furthermore, all of the pixels we want to remove belong only to complete particles, so this tells us we should focus on the anti-extensive particle operations which use conditional dilation. After this, we need a way to distinguish the particles we want. Size (maximum width) and length immediately come to mind, but an examination of the image shows that these won't work here. One thing that does distinguish the disks from the polygons is shape, alternatively expressed as smoothness or roughness, or as the presence or absence of corners. Our knowledge of the effects of the opening residue tells us that we can use it to separate these two classes of objects.

As a final example, suppose that instead of wanting to get the disks in an image like Figure 3.2, we change the problem such that the accept mask includes only the large disks, and the triangle (which has sharper angles and thus greater residues than the squares). In this case the algorithm is still anti-extensive, it still has to get whole particles, but now it has to use a combination or disjunction of conditions: one of size, and one concerning shape or residues. One possible solution to such a problem, and a series of steps or algorithm transformations that one might go through to obtain it, is illustrated as a sequence of algorithm graphs, in Figure 3.3. This sequence was developed without using the notion of band operators described in the last chapter.

I have presented these fairly simple problems with easily describable solutions for the purposes of exposition—naturally, if things were always so clean one could generate the solutions simply from the problem descriptions

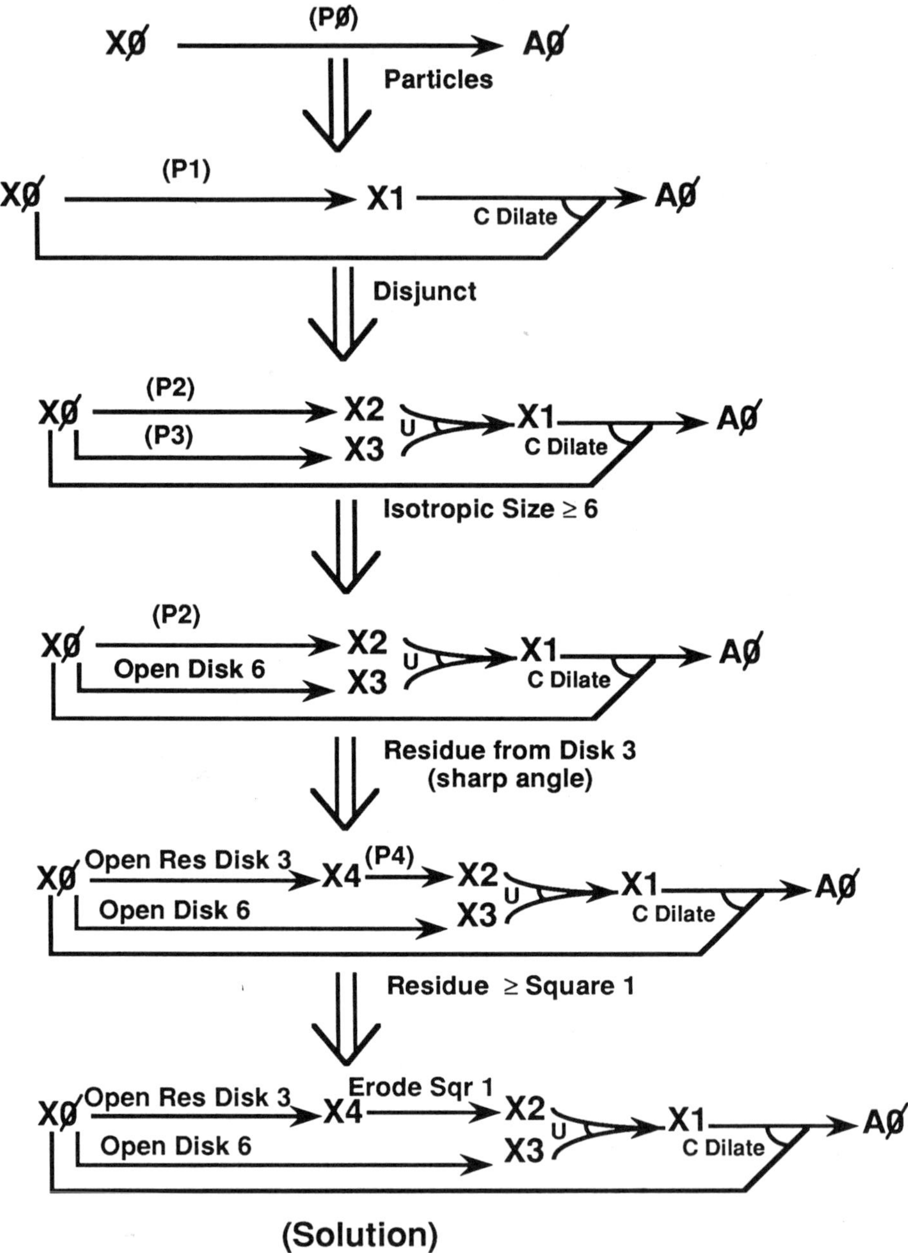

FIGURE 3.3. Example Algorithm Transformation Sequence

as I have done here. In general, the problems are not so clean. Sometimes it is not possible to give a simple English description of the desired solution; often the images are noisy, and the desired result is not necessarily one of the connected components in the original image. This is what motivates the use of the mask images to describe the goal of the recognition algorithm—a human can select the ideal result for each example image, by drawing it, without having to come up with a description that may be difficult to generate or formalize. It is the machine's task to then come up with a mathematical description of the desired set, based on the constraints imposed by the masks. A more complete system would also be able to take in additional known constraints expressed in a restricted natural language, but for the present I have not included this possibility.

3.3 Formal Description of Problems

In this section we want to take a more formal look at the kind of problems we are attempting to solve here, to get both an idea of why these problems are hard, and what we are up against when we try to solve them either automatically or by hand. First, to get an idea of the enormity of the numbers with which we are dealing, suppose we have a binary image X with a spatial support of MxN = P pixels. This image X is drawn from the space $\mathcal{X}$, of all binary images with the same support, numbering some 2^P in total. (For the very small 64x64 images I have used in the implementation, P is 4096 so we are talking about 2^{4096} possible images).[3] An arbitrary image-to-image function is a mapping on $\mathcal{X} \times \mathcal{X}$, that is, for each possible input image in $\mathcal{X}$, we choose one output image also from $\mathcal{X}$—this gives 2^{2P} possible image functions, or 2^{8192} over just our small 64x64 images. For a given recognition problem to be solved, we can consider as a first approximation that we are looking for one specific image function among this huge number that gives the desired mapping from input images to output images.

There are of course some qualifications to be made here. First, for a given problem the set of possible input images it considers may be a small subset of $\mathcal{X}$; likewise, the set of outputs is likely to be an even smaller, perhaps different, subset of $\mathcal{X}$. In addition, we often don't have to have exactly perfect results over all images in the domain of the desired function, meaning that a number of the possible functions would serve equally well as solutions. Finally, the large majority ('nearly all') of the possible image functions are chaotic, in that they may produce wildly different results for

[3]Likewise for a 512x512 8-bit grey level image, more than $2^{2,000,000}$—just think of the possiblities!

some very similar inputs. In practical situations we are not interested in such functions, and the image domains that we are interested in are also generally more 'well-behaved' than the average image in $\mathcal{X}$.

While all these points are quite valid, they do not compensate very much for the fact that the number of algorithm tools or operators we have available to try to construct the desired mappings or functions, even when we consider all parameter settings and so on, is probably on the order of millions at most, and even by chaining together sequences of them to produce more elaborate functions, we are nowhere near the cardinality of the set of possible functions. In short, compared to the enormity of possible image functions that exist as distinct mappings between a set of input images and a set of desired outputs, our tools for finding and constructing such functions are rather meager. As algorithm developers, our hope is that the image sets and desired results we work with are really very well-behaved, so that there is some possibility of reaching reasonable solutions in a fairly small number of steps using the tools that we have.

The situation is actually worse than I have described it here and in the introduction, where I couched the set recognition problem in terms of trying to find the right accept set or A_i, for every input image X_i. In reality, this is an oversimplified view, which is true only when the reject set for each image, R_i, is just the complement of A_i. In many cases that is not true; as stated before, the union of the disjoint A sets and R sets for each example image, may not cover the entire image space. This is particularly the case as we try to solve subproblems and sub-subproblems of the original problem. Other areas in the images outside of A and R are simply to be ignored.

What this additional element of variability means is that for a given input X_i there may be many more acceptable outputs which satisfy the twin conditions that they contain all of A_i and none of R_i. In effect this greatly increases the total number of possible image functions given above for the binary case, since now there is not just a single valid output associated with each input X, for a given image function, but rather a *set* of such outputs, each of which must lie within the bounds dictated by A_i and R_i. The number of ways to choose two disjoint sets or masks from $\mathcal{X}$ to bound the acceptable outputs is huge, compared to the single choice we have for each X if there is only one acceptable output, and the total number of possible image functions is based on the cross product of this larger set of allowable output choices with the set of input images.

A better way to look at a set recognition problem or subproblem is to think of it in terms of a set of triplets, $[\mathrm{X}/\mathrm{A}/\mathrm{R}]_i$, $i = 1 \ldots N$, where X is drawn from $\mathcal{X}$, the space of base images, A is drawn from $\mathcal{A}$, the space of binary images composing the accept masks, and R is drawn from $\mathcal{R}$, the space of binary images composing the reject masks, with a major restriction being that A and R never intersect, for a given i. We can think of a *problem space* as a subset of the cross-product of $\mathcal{X}$, $\mathcal{A}$, and $\mathcal{R}$, such that each X_i appears no more than once, and such that for all i, A_i and R_i are disjoint. Individual *instances* of problems can then be thought of sample sets taken from this problem space. Different problems will have different combinations of triplets which make up this space. When we choose individual example base images, along with their associated accept and reject masks, we are in essence taking a sample set from a particular problem space, and trying to develop an algorithm on this set which will correctly apply to any other problems drawn from the *same population*, as represented by that problem space.

This concept is very important so let me reiterate: When we select a set of example images along with their accept and reject masks as instances of a problem or subproblem, we should think of it as selecting samples from a problem space which is a particular *relation*, or subset of triples, from the space $\mathcal{X} \times \mathcal{A} \times \mathcal{R}$, with the conditions that each X appears only once and the associated A and R sets for each X are disjoint. The extent to which we are able to solve the problem for the entire space, based on the sample we have chosen to work with, will of course depend on the extent to which that sample reflects the major important characteristics of the other instances in the problem space, i. e., on the extent to which it is an unbiased sample from that space.

For any given set discrimination problem or subproblem, there is also very often a fourth element beyond the set of base images $X_{1 \rightarrow N}$, the set of accept masks $A_{1 \rightarrow N}$, and the set of reject masks $R_{1 \rightarrow N}$, which helps to define or constrain it. The fourth, if present, is any other knowledge which one may have about the problem which may help to constrain the search. This may be information which is ultimately derivable from the images themselves, or it may involve knowledge of the problem domain and context that would be very hard to derive empirically. In the implementation effort described here, this knowledge includes also predicate information about the relationships of the X, A, and R image sets, which is computed at one time, and later passed on from a parent problem to one of its subproblems, in an attempt to make it easier or faster to solve that subproblem. The importance of such additional knowledge is that it can help to constrain the potential search space for an algorithm, which may be especially needed if the original sample set of images chosen is not large enough or representative enough to adequately do so.[4]

In order for a problem to be what I will call a *meaningful* one, all of the problem instances in the space should be such that the same key set of

[4]In what follows, I will usually speak of a subproblem being defined in terms of the X/A/R image sets, but it should be kept in mind that in some cases, there may also be additional knowledge present which helps to constrain it beyond what is given by these images.

image criteria universally distinguish the accept set pixels from the reject set ones, at least to the level of accuracy desired. If this is not true then only an 'enumerative' solution is possible for the problem space.[5] If it is true, then it may be possible to find an algorithm which 'solves' the problem represented by the space, *if* some set of operators or sequence of them exists which address the key criteria of the space, *if* one can find out what the key criteria are by observation, *if* one can make the association between these criteria and the operators which address them, and *if* one can avoid focussing on irrelevant criteria or missing important criteria because the sample chosen was somehow biased or simply too small. In short, even if the problem is a meaningful one and specific image criteria exist which can solve it, the problem of solving it may not be easy or straightforward.

In the problems discussed in the system implementation described later, I am assuming that there is always a set of meaningful criteria which solve them. This may not always be the case for real world problems; we often cannot know for sure. In general we have no guarantees for real problems that such criteria exist—except to the extext that our own human visual abilities tell us that they must.

The REM system that I have implemented has been designed to handle certain specific image criteria which were summarized in Section 2.16. While the system has solved a number of problems using a only a few different base images, (in order to make the differences in the problems more clear), the techniques it uses are entirely general, depending only on the *relationships* found between the X/A/R sets of each sample instance, and in no way on the particular images used. My claim is that the REM system will solve *any* problem for which a solution exists in terms of one (only) of the image criteria listed in Section 2.16. That is, one can choose any other sample of particular image triplets making up a set of problem instances, and as long as the problem space from which these were chosen has a meaningful solution whose key criterion is one of those chosen from the set listed, then the REM system will find that solution, unless the sample chosen is biased in such a way that a simpler solution exists for those particular instances drawn. Some of the reasons why this is true are presented later in Section 3.6, others will be found in the explanations and discussions found in Chapters 4 and 5.

Before closing this section, I want to explain why there can be no trivial solutions to the problems solved by the REM system as described in Chapter 4. By trivial solution, I mean one which has the character of a table look-up—that is, one where the program would be given the input image

[5] An *enumerative* solution is simply a complete listing of the mapping of input images to correct output images. The possibility that more than one key set of criteria are needed to cover the problem space, because of distinct differences in the images, I would consider here as a single key set of *disjunctive* criteria.

instances, do a simple look-up and print out the correct algorithm, if one existed, from the many thousands of specific algorithms that it supports. The main reason there can be no such solution to the problem, (aside from the fact that the stepwise algorithm construction process and algebraic language used here preclude such methods), is because the universe simply isn't big enough.

Since each problem is expressed as a set of a few triplets of X0/A0/R0 images, for the system to solve these problems by table look-up, it would have to recognize that the triplets composing each instance in the sample, *each* constituted an example that could be solved by the *same one* of its algorithms. In other words, for each of the thousands of specific algorithms it supports, it would need to know *all* of the X0/A0/R0 triplets which could be solved by that algorithm, since it could never know in advance which combination of triplets it might be faced with, nor even whether they were all examples of the same algorithm, or different ones, or not examples of soluble problems at all. The look-up would require searching through all of the thousands of tables (one for each supported algorithm), to see if all of the triplets in the sample set existed somewhere on the same *one* table.

For the problems solved by the current system, all base images (X0) from the space $\mathcal{X}$ are allowable, which means that there are 2^P (2^{4096}) allowable triplets for each specific algorithm supported, only in the case when $R0_i = A0_i{}^c$, and many more if this is not the case. Thus, even if only one algorithm was supported, the task of being sure, by table look-up, that the set of three or four example instances chosen for a given problem were all examples that could be solved by that algorithm, would require far more bits to store the tables of triplets than there are elementary particles in the visible universe.

3.4 Algorithms as Graphs

As the purpose of algorithm development in general, and the REM system developed here, is to produce algorithms, the representation of algorithms is a critical part of the overall design. Indeed, it is important not only to be able to represent complete algorithms, but also algorithm *states*—that is, algorithms which are only in a partial state of completion. While a random or exhaustive search technique would only have to know how to generate completed algorithms, a system which tries to develop them by stepwise refinement needs to have ways to represent partial solutions and to package what remains into 'subproblems' to be solved later.

All of the operands used by the REM system implementation are fixed-size images, and all of the operators are image-to-image ones which take one or two input images and produce one output image. Because of this,

Initial Algorithm State

Incomplete Algorithm State

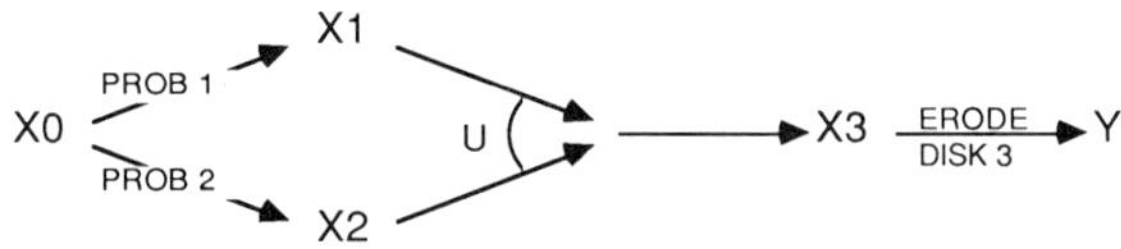

Terminal Algorithm State

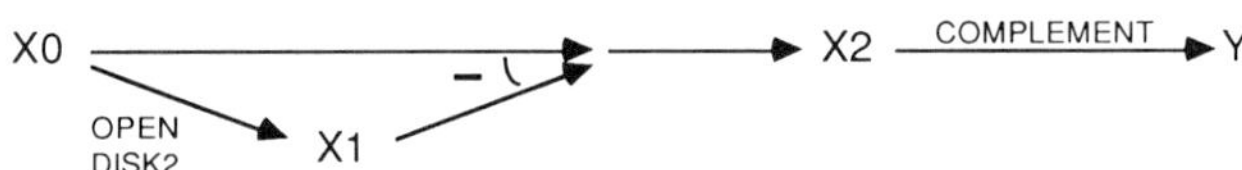

FIGURE 3.4. Algorithm State Graph Types

algorithms can be conveniently represented as directed graphs having one or two source nodes and one destination node. The nodes represent images while the labeled, directed arcs represent image operators. (Combination operators form a special kind of two-into-one directed arc—see Figure 3.4). In this directed-graph representation, complete or *terminal* algorithms have all of their arcs labeled with an operator. Incomplete algorithm states will have one or more arcs labeled with a subproblem ID, indicating that something needs to be filled in. The initial algorithm state when the system starts up consists of a source node (the input image), a destination node (the output image), and a single arc between them labeled with the overall problem ID (PROB0), which represents the entire problem to be solved. This *null algorithm* is shown at the top of Figure 3.4.[6]

---

[6]It would be possible of course, to have a system like this alternatively start with a more complicated algorithm state, representing a partially solved problem. Its job would then be to complete the algorithm given this framework with which to begin.

Algorithm states are modified by replacing a subproblem arc with something else; such modifications are called *program transformations* (see Figures 3.3 and 3.5). One possibility (the ideal case) is to simply fill the subproblem arc in by labeling it with an operator which completely solves that subproblem. Another is to replace it with a sequence of two arcs and a new node between them—one arc is labeled with an operator, representing a partial solution, and the other is labeled with a new subproblem ID, representing the part that remains to be solved. The new node is a new intermediate image result in the overall algorithm, which will need its own image ID. A third possibility (see Figure 3.5) is to replace the arc with a combination operator, producing two new arcs and two new nodes (intermediate images). One or both of the new arcs may be labeled with new subproblem IDs, representing new subproblems to be solved; in some cases one of them will already have a solution given to that half of the problem. Figure 3.3 showed a possible sequence of algorithm state graphs from the null algorithm to the terminal algorithm which solves one of the example problems described in Section 3.2. The program transformations can be inferred by watching how the graph changes from one state to the next.

Since new algorithm states or graphs are derived by transforming old ones, the set of algorithm states examined by the program forms a tree, with the initial one-arc null algorithm state described above at the root (ALG0). This *Algorithm State Tree*, or 'AST', has an algorithm state description at every node, along with additional information about how well it performs, its associated costs, and the subproblem and image IDs it contains. An example AST also is represented in abstract form by Figure 3.3. While this example only shows the algorithm states leading to the final result, a real Algorithm State Tree might have other branches leading to other algorithm states or graphs that did not develop into solutions. Terminal algorithms on the Algorithm State Tree appear as leaf nodes which have no subproblems—they are not the only leaf nodes on the AST since typically some nodes will not have been expanded out completely at any given time, even when the search is terminated.

In a similar way each subproblem is derived from a previous larger problem, and thus these too form a tree, the *Sub-Problem Tree*, or 'SPT', which is used to keep track of problems created and solutions examined. Each node of this tree contains a description of a subproblem (X/A/R and any additional knowledge), with the original problem to be solved represented by the root node (PROB0). Instead of storing the base images and masks for each subproblem as its description, (which would take up too much space), these images and masks are expressed in terms of how to calculate them from the base images and masks of the original problem. Along with the subproblem description, each node also has a list of algorithm states in which the subproblem appears, a 'blackboard' of facts (predicate relations) which are true of the subproblem, and test and evaluation data for the

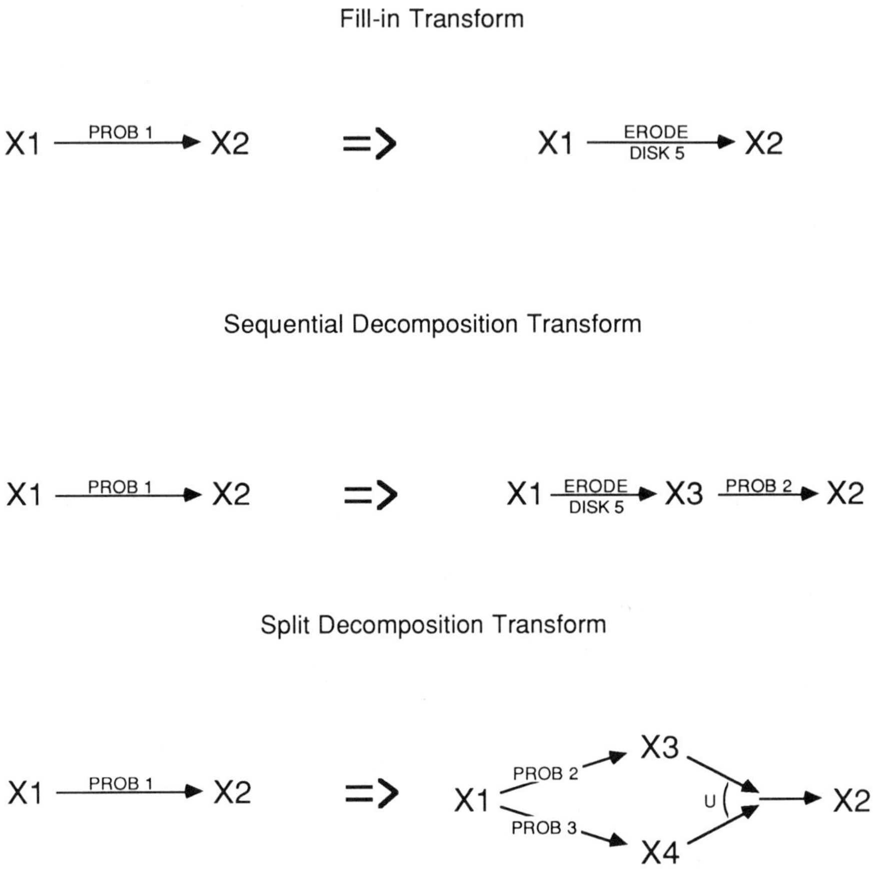

Fill-in Transform

X1 —PROB 1→ X2 => X1 —ERODE / DISK 5→ X2

Sequential Decomposition Transform

X1 —PROB 1→ X2 => X1 —ERODE / DISK 5→ X3 —PROB 2→ X2

Split Decomposition Transform

X1 —PROB 1→ X2 => X1 —PROB 2→ X3, X1 —PROB 3→ X4, X3 U X4 → X2

FIGURE 3.5. Types of Program Transformations

operators proposed as solutions to it.

The Algorithm State Tree and the Sub-Problem Tree are separate entities for the sake of efficiency. The same subproblem may appear on more than one algorithm state, and the same algorithm state may contain many subproblems; this separation lets us describe each only once. Furthermore the two trees are updated differently. Replacing a subproblem arc with with an operator generates a new algorithm state, but not a new subproblem. Likewise, some algorithm transformations produce two new subproblems, which have to be treated as a connected pair, but only one new algorithm state.

Termination of the overall search process as well as the construction of these two trees will in general depend on the requirements of the problem—in some cases it may be when the first algorithm that succeeds is found. In other cases, termination may depend on the overall time spent, the accuracy of the solution found, or the cost in time and space of the algorithm as measured by some external evaluation function. The implemented system described in Chapter 4 stops when it first finds a satisfactory solution.

That completes our overview of the representation of algorithms used in the solution of set recognition problems. The concept of representing algorithms as graphs and developing algorithms progressively by making tranformations of those graphs is a very powerful one which is commonly used in other types of automatic programming systems. The graphs themselves are isomorphic to algebraic representations of the algorithms in which the nodes are represented by the left hand sides of equations, and the arcs are represented by the algebraic image expressions or operations which label them (right hand sides). More details on this subject are presented in Section 4.1.

In the next section, we are going to look at the process of algorithm development, in order to understand the typical sequence of events carried out by human algorithm developers, and to understand the major important components and problems engendered by this process.

3.5 The Algorithm Development Process

Now that we have some idea of how algorithm states, subproblems, and their associated trees can be represented, I want to examine what are some of the elements of the image algorithm development process, how they fit together, and where the real difficulties lie. Typically one starts out with a set of example images, all containing some objects or sets of pixels to find, or *recognize*. The first thing that has to be done normally is to catalog the example images in some kind of a database, and then to *truth* them, by having someone identify those regions of pixels on each image which are the goal of the recognition algorithm. In most cases all other pixels are to be eliminated, but sometimes it is only important to eliminate some structures, while others may be of no concern one way or the other. These masks, representing the pixels to accept and to reject for each original image, are stored along with them.

When the cataloging and truthing activities are finished, the actual algorithm development process can begin. Assuming no a priori knowledge about constraints on the problem or the images, one starts by looking at the original images and masks for image criteria, (relationships between objects or pixels in the image), which clearly and consistently discriminate the desired regions from the undesired ones. Visual observation is generally used to come up with possible discriminating criteria, and then on the basis of one's knowledge of the field regarding the available operators and how they affect images, certain operations can be selected which address the criteria under consideration.

At this point one begins to test the operators chosen on the example images in an attempt to find out whether, and/or how well they will work. Part of this process involves selecting the ideal parameters for a given operator, a task which can be tricky because one has to evaluate the various possibilities over an entire image set, and come up with some average notion of the goodness and consistency of performance, compared to other alternatives. Sometimes data is collected on the effects of an operator over the sample images for different parameters, but quite often these evaluations are made simply by trying the operation on several images, and integrating the results visually or mentally.

Once one has tested some candidate operators, the next step is to choose the best alternatives, and then go about trying to solve any subproblems which remain, again by first observing the images carefully, selecting new criteria to perform the remaining discriminations, testing the operators which seem appropriate, and then evaluating and choosing between them. This kind of iterative cycle continues until a satisfactory solution has been reached for the overall problem, or until one gives up on the problem as being unsolvable within the existing constraints on resources and time.

Having completed this brief overview of the algorithm development process as a whole, let's now look in more detail for a moment at what are the essential components of this process and where the problems, errors, and bottlenecks tend to occur. Some of these components concern the representation of things, some concern collection and management of data, some concern ways to evaluate alternatives and make decisions about them, and still others concern the knowledge needed to perform all of these activities.

First of all, one needs a way to represent both partial and complete algorithm states, and to execute algorithms or individual operations. This topic was discussed in the previous section; in general this representation is based on whatever language is used to drive the image processing hardware. In the system implemented here, four levels of representation were used: A target language made up of band operator equations; an algebraic language level for representing lower level image operations in a way which also supports the graph representation described in the previous section, using a minimal number of primitives; a Lisp command level which formed the executable interface to the image processing system; and finally the image processing language normally used to program the hardware. Partial algorithms are a little more difficult to represent than complete ones; for human algorithm developers they often exist only as notes on scraps of paper, scribbled by the algorithm developer while working. In the system implemented here, they are supported only at the target language level as arcs on a graph or image expressions which are replaced by the name of a subproblem.

One of the other major components of the algorithm development process is the collection and management of data of various types. For example, as the development process proceeds, there are often several incomplete algorithm states (lines of attack) that are being pursued at one time in more or less parallel fashion. Keeping track of all of the states and what one is doing on any given state can be a major mental burden. Likewise, as the development process continues, new and different kinds of subproblems are created; one has to remember them, remember how they are related to other subproblems and algorithm states, have ways of performing the intermediate calculations which lead up to them, remember which portions of the images were to be discriminated as the goal of each of them, remember what operations one was trying on them and what were the results so far, etc.This too can become a tremendous mental burden, simply from the data management point of view—one which can easily lead to errors.

In the REM system developed here, algorithm states, subproblems, and the relationships between them are managed by means of the Algorithm State Tree and the Sub-Problem Tree, both described in the previous section. An Agenda is used to maintain a prioritized list of the subproblems currently being pursued. The ability to to generate the intermediate images, and perhaps more important, the specific (modified) accept and reject masks which together define each subproblem, is achieved by maintaining algorithms with each subproblem object for calculating them from the original images and masks of the overall problem. Likewise, the list of operators being tried for a given subproblem are also maintained as part of the subproblem object.

Another data management component concerns how one keeps track of the results of all of the tests being performed with different operators and different parameter settings on a given subproblem, especially when it may take some time to execute them on the entire example set. If this data is being generated by a computer, this is less of a problem, but quite often this kind of testing is done entirely interactively, relying on the eyes and mind of the algorithm developer to store and integrate them, which again can lead to errors and biases simply because of the time and number of images and calculations involved. In the system described here, all of this testing and the associated analysis is done by the program, using routines specially developed for each class of operator. The raw results and the summarized results of the analyses are also placed in slots of the given subproblem object that one is currently trying to solve.

Beyond the problems of data management, one has to have appropriate means for evaluating operators and subproblems, and for making decisions. For example, one needs to be able to evaluate new subproblems, in order to decide how to solve them. One has to be able to evaluate how well a given operator being tested solves a subproblem over the entire set of example images, as well as how consistent it is in doing so, and even how costly it is. One also has to be able to evaluate entire partial algorithms, in order to decide which ones look most promising. Most of these evaluations are handled entirely visually today by algorithm developers working interactively. This is not all bad, since the eyes and brain can do a very good job at recognizing gross differences which suggest rejecting certain alternatives, rather quickly. In the cases where there are finer, less consistent differences, however, this practice can lead to certain kinds of errors and biases.

In the REM system, there are mathematical routines for initially evaluating subproblems in terms of how well they match the desired accept and reject sets; this provides some information for deciding which strategy to use in solving them. The ability of a given operator to solve a subproblem is also measured by similar techniques, and summary measures of accuracy and consistency of these solutions over the entire image set have also been developed. In the system's current state, not much is needed in the way of evaluating entire partial algorithms (since most problems have fairly direct solutions), but the measures used here would be similar to those mentioned above, except that they would be based on the effect of the entire partial algorithm. One way to do this would be to do a piecewise analysis of each subproblem remaining to be solved; another would be to run the partial algorithm by replacing subproblems with identity operations and looking at the overall result.

In the area of decision making, one has to have the means for choosing between alternative solutions to subproblems, for deciding which subproblems to try to solve next (prioritization and resource allocation), and for deciding what are the optimal parameters to use with a given operator over the entire image set. All of these are rather important, since making a mistake may mean that you throw away what would have been a good solution or direction, or that you waste a good deal of time solving problems that were ultimately not important because they didn't lead to a solution.

These particular elements are not elaborate in the current system simply because the problems solved at present are fairly straightforward ones, which don't generate a lot of alternatives. As one begins to look at more complex problems, however, where a single operator can only partially solve a problem, where many operators may be able to offer partial solutions, where there are multiple ways to select the parameters which partially solve the problem, multiple strategies which can be pursued and multiple ways to generate subproblems—then these issues of prioritization, resource allocation, and decision making will become much more important. Human algorithm developers often proceed according to whatever seems good at the moment, which may be correct or which may be influenced by irrelevant factors. This is an area where more research needs to be done, but I believe that there are good mathematical priciples which can be developed to improve the decisions made in these situations as well.

Finally, the algorithm development process also requires having different kinds of knowledge available. Most of the required knowledge is about the image operators—their properties, how they work, when they are appropriate, how to collect data on their effects which will allow you to select or discriminate pixels, and perhaps most important, what image criteria they address. Other knowledge concerns strategic issues of algorithm development such as how to decide what direction to go in based on the current state of the algorithm and the types of subproblems left to solve. Finally, one has to have knowledge about the different ways of generating subproblems (sequential vs. parallel decomposition), and when each is appropriate. This also includes knowing how constraints on the search required for the parent problem can be deduced, modified, and passed on to subproblems generated from it, so as to reduce the amount of work needed to solve the new problems.

In the REM system, this knowledge is distributed in different places. Some of the knowledge about operators, how they are defined and how to collect data for them is collected into an operator data base, and an algorithm base of data collection routines. Information about when they are appropriate to be used is contained in the operator expert which makes decisions about which operators to propose in a given situation. Knowledge about ways of generating subproblems, strategies to use in attacking a given one, and how to analyze data and choose parameters of operators toward these ends, are contained in the analyzer part of the code, which has to offer candidate solutions to subproblems. Other strategic knowledge concerning choosing between alternative solutions, prioritization of activities, and so on, is contained in the routines responsible for these decisions.

All of these elements are part of the algorithm development process, and it is necessary, in developing any automatic means of doing the same thing, to have all of these bases covered in one way or another in the code, just to make such a program work at all. One important difference between automatic systems and human algorithm developers, in this regard, is that humans have eyes and brains which machines do not, which allows them to make assessments of operators, make decisions, and decide what is appropriate intuitively, without conscious thought. In the mechanized systems we have to replace these functions with measurements or calculations that in the end may be more consistent and reliable, but which do not have the same character or flexibility that goes with human vision.

That completes our overview of the algorithm development process, and its major components. The implementation of each of these components in the REM system is described in detail in the individual sections of Chapter 4. In the following section we want to look at some of the methods used by this system to reduce its search, and to show why these methods will still not miss any valid solutions.

3.6 Search Strategies and Completeness

In the last section, we looked at some of the important elements of the algorithm development process. One of the most important of those elements is the search through the space of possible algorithms. In this section, we want to look at the question of how one can do an intelligent search through this space; in particular, what knowledge can one use to speed up this search, or avoid examining all possibilities one at a time, and yet be sure that one is not going to inadvertently throw away possible solutions. The purpose of much of what will be said in this section will again be to give an overview of the issues and the principles involved. More detail will be found on these subjects when we get into the description of the implementation (Chapter 4), and the results achieved (Chapter 5, especially Section 5.4).

One of the most important aspects of the morphological operators being used here is that they come in pairs, one which works on the foreground (the anti-extensive operators), and the other which works on the background (the extensive operators). These basic operators, and the band operators which are derived from them, are thus especially suited to situations where a strict containment relationship exists between the base images of a problem, and either the accept or reject masks (or both), over all of the image instances. There are four main possibilities to consider.

In the first case, called an *A-Sup* condition, the base image X is always an upper bound on the accept set A, or in other words, X always contains *all* of A, so that we just have to eliminate the part of X which has pixels that we don't want. This situation suggests the use of an anti-extensive operator, such as the erosion, opening, or infmasks. In the opposite or dual case, called an *A-Inf* condition, the base image X is always a lower bound on the accept set A, which means that it contains *only* A pixels, and none from the reject set R. (Whether it also contains pixels from ignored regions is immaterial). This case suggests the use of an extensive operator, such as the dilation, closing, or supmasks. The A-Sup and A-Inf conditions are the preferred kinds of problems for the REM system to work with, since in these cases it can apply the main operators it has directly.

Two other conditions which are nearly equivalent to the A-Sup and A-Inf conditions, are the *R-Sup* and *R-Inf* ones. The R-Sup condition occurs when X is always an upper bound on the reject set R, that is, when X always contains all of R. In this case if we complement X we will automatically have an A-Inf condition, which can be solved by an extensive approach. Likewise, the R-Inf condition, which occurs when X always contains only R (none of A), can be turned into an A-Sup condition by complementing, which is then amenable to the anti-extensive approach.

These four conditions really form the bedrock of the strategies for solving set recognition problems based on morphological operations. The overall problem is completely solved when both A-Sup and A-Inf hold, meaning that X is both an upper bound and a lower bound on the accept set A, or in fact equal to it. Likewise if both R-Sup and R-Inf are true, then X equals the reject set, so the overall problem is solved simply by complementing.

Sometimes none of these four conditions hold but we are close to one of them. For example, we might have a situation which was almost an A-Sup except that a few pixels in the accept mask sometimes were in the background of the base image, rather than the foreground. If the percentage of pixels involved is small enough, it might be acceptable to ignore them. This could be done by modifying the accept masks in each instance to contain only pixels in X (the foreground). Similar approaches can resolve small mask errors in the other three cases. Another situation occurs when we have an A-Sup or A-Inf condition, but when we analyze a particular operator, comparing the distributions produced for the accept sets versus those produced for the reject sets, there is always a slight overlap in the ranges. If this is small enough, we could simply ignore it by modifying the distributions for one of the two sets, which would create disjoint ranges for the resulting distributions, and which is what our band operators require.

If the overlap in ranges is too large, or if the two sets actually share many of the same values under the operator being examined, then we may have to figure out a way to select a partial solution and generate a new subproblem, especially if no other operator seems to offer more hope of a complete solution. In generating a subproblem, however, we would prefer to choose our parameters, if we can, so that the type of subproblem will be one of the four (A-Sup, A-Inf, R-Sup, or R-Inf), which will allow us to attack it directly with one of our previously described strategies, based on strict containment. If this is not the case, perhaps because the two distributions overlap too much, the best approach may be to try and select parameters which produce the minimum total error (sum of false alarms and misses) between the two sets. The problem with that kind of solution is that we will have to generate *two* new subproblems to solve—one to eliminate pixels we don't want, and the other to pick up the ones which are missing. If we have to do that too many times in succession, our Sub-Problem Tree will start growing exponentially, so this situation is to be avoided.[7]

Whenever we have to generate new subproblems, there are often several alternatives from which to choose. In addition to those of trying to choose parameters which leave an A-Sup or A-Inf condition, or which minimize the error between the two sets, there is also the possibility of dividing the original problem into two parts, to be solved separately and combined with a union or intersection (parallel or split decomposition), instead of accepting the partial solution found and then creating a subproblem for an operator which is to follow (sequential decomposition). The value of one method versus the other depends on whether it is easier to solve the remaining problem based on the starting image of the original subproblem, or based on the image which resulted from the partial solution operator.

In addition, if we are in a case with a partial solution, and there exist ignore regions (areas not covered by the accept or reject sets), then in setting up the subproblems we can choose to concentrate either on trying to 'capture' the accept set *or* the reject set. (If there are no ignore regions, then any algorithm which determines either the accept set or the reject set by definition also determines the other, as they differ only by a complement. The duality of the operators used then dictates that one should concentrate only on obtaining the accept set). When ignore regions do exist, however, there may be different algorithms which are better at selecting either the accept set or the reject set, or which can do so at less cost, where cost is defined in terms of the number of elementary steps required to execute the algorithm in hardware (see Section 2.9). Either approach solves the problem, because the goal is just to find a way to distinguish the two sets, and even if we get the R set in the foreground, we can just complement it to get A plus some of the ignore regions which we don't care about. In these cases one has to then consider both strategies. When the given problem or subproblem can be solved in one step, this doesn't make a difference, but if multiple steps are required, then the strategies to get the A set versus the R set will differ in how they generate subproblems.[8]

Now that we have looked at some of the strategies for solving morpho-

---

[7]While there are some morphological operations such as smoothings which both add and delete pixels, their properties are less well understood, and it is more difficult to build automatic selection criteria for them than for those which obey the inclusion properties.

[8]The strategies described in these last three paragraphs have not yet been implemented in the REM system, but hopefully will be at a later date.

logical set recognition problems, I want to next turn to the issue of what are some of the ways that we can reduce the search for algorithms, especially without creating the possibility of throwing out actual solutions by accident. The REM system, as implemented at present, is in a sense a breadth-first exhaustive search for an algorithm, which stops whenever it finds an acceptable solution. Stated more clearly perhaps, it can be *nothing but* a breadth-first exhaustive search *unless* some intelligence is added so that it can avoid work which isn't really needed. Since I want it to always be able to find a solution if there is one, it has to in some way deal with *all* of the space, because if it skips over certain parts, it risks to miss the solution. The only way it can avoid to do the work of an exhaustive search, and yet make the same guarantees about finding a solution if there is one, is to find ways to reject large sections of the search space all at once, on the basis of comparatively simple tests, rather than to reject each incorrect algorithm one-by-one, as the exhaustive method does.

The basic philosophy used here to reduce the search, then, is to find relatively inexpensive tests which reject a class of operators only if the whole class cannot possibly provide a solution to the problem. Alternatively when we test for whether a particular class of operators should be considered, we use a test which places a *necessary* condition on having a member of that class offer a solution. In other words, if the test is passed then there is a possibility that a member of the class may offer a solution (especially if it succeeds on several image instances), but if the test fails then there is no possibility of such a solution. The efficiency requirement for each such test is that it requires significantly less computation than the analysis or individual testing of algorithms that it replaces, in the case when it does actually reject the whole class, and that the probability of it succeeding if there is no solution is relatively low. Moreover, tests of this type which are especially good will be those which are so *specific* to the class of operators they are about, that if they succeed, it will be unlikely that other tests used to propose other classes of operators will also succeed.

A few examples will illustrate the point. First, the band operators used by the REM system require disjoint ranges of values of the accept and reject sets, in order to guarantee a complete solution. Testing for disjoint *ranges*, however, is more difficult than testing to see if the distributions themselves are disjoint, which is a necessary condition. If the distributions themselves are not disjoint, then there is no point looking for disjoint ranges which will provide a complete solution: There won't be any.

As another example, we do our tests for the four containment conditions, A-Sup, A-Inf, etc. If true, they will tell us whether to select from the group of anti-extensive or extensive operators. These are necessary conditions; if one of the extensive or anti-extensive operators is to solve the given problem completely in one step, then the associated one of these conditions must be true. The test performed requires that a simple measurement be done on each of the input instances, which if false rejects roughly half or a quarter of the algorithm space (and all the analysis and/or testing of individual operations which would go with this), without any possibility of throwing away the correct solution, under the assumption of a one-step solution.

For additional cases, let's look at how we choose between more specific classes of operators. For example, if we find that for all of the particles (or holes) of X we either have the entire particle in A or else in R, this suggests that we should consider the particle (respectively, hole) operators. It doesn't guarantee that we will find a solution from this class, but if this isn't true, then these operations *can't* provide a complete solution, so rejecting them does not pose a danger of throwing away the solution. This test is much shorter than testing the whole class of these operators for a solution—it requires simply testing the image instances to see if no particles are partly in the A set and partly in the R set. Moreover, if this test is true, then it can also be used to eliminate other classes of operators, such as the erosions, for which it could never be true.

As another example, one effect of the erosions by compact (non-linear) structuring elements on the foreground, or dilations on the background, is that they change the entire border. Openings and closings by compact structuring elements, however, leave parts of the border intact—that is just the nature of what they do to the image. By testing the borders of particles or holes to see if the borders are entirely within either the A or R set, or mixed between them, we can distinguish between these two classes of operators, and thereby eliminate one of them. Since finding the border requires using only the smallest structuring elements, this test is fairly fast, much faster than the analyses that would be required on the whole class being eliminated. Likewise, there are ways to distinguish between erosions and dilations, and openings and closings, by squares versus octagons and diamonds, by looking at the effects on the image of using the smallest square and diamond-shaped structuring elements, and comparing these with the accept and reject sets.

As a final example, the test used to propose the position masking operations (infmasks and supmasks), is based on checking to see if there is significant overlap in the addresses of the pixels belonging to the foreground or background which are in the A and R sets, respectively, along the X or Y dimensions. If there is no significant overlap when we look at them individually, then there may not be when we look at all of the instances together. If there ever is a significant overlap, then this kind of operator is not going to offer a complete solution. Once again, the predicate test is simpler than the analysis it replaces over the class of operators being considered.

In all of these cases I have chosen tests which are cheap compared to the analyses they replace; which, if false, correctly reject a whole class of operators without question of missing a solution; which, if true, suggest that an operator within the class may provide a solution; and which in many cases, if true, will help to eliminate other classes by virtue of their specificity. Whether tests with these characteristics can be found for cases which do not admit complete or perfect solutions, is another very good question. I have not yet been able to investigate this thoroughly; my sense is that there may be such tests in certain cases, but they will be more difficult to find, and will undoubtably be less specific. The step from problems which can essentially be solved directly by one band operator, to those which necessitate more operations, is a big one, and we may find out later that in order to do a reasonable job of covering that vastly larger space, we may have to depend on heuristics which do not guarantee finding an existing solution in all cases.

Before closing this section and this chapter, I want to also mention some of the other methods or themes that can be used to reduce the amount of work that has to be done in solving these kinds of problems. One of the most important, which perhaps should have been mentioned above, is that it is possible to *analyze* the effect of operators rather than testing all combinations of their parameters. The monotonicity properties discussed in Chapter 2 which allow us to construct distributions for the extensive and anti-extensive morphological operations, much like the histogram for grey level images, make it possible to analyze all possible parameter combinations simultaneously, and is entirely equivalent but much faster than testing all pairs to see if they solve the problem. This is where the REM system gets its greatest reductions in speed. (See Section 5.4 for more discussion on this subject).

Another way to reduce calculations is to perform initial tests when proposing possible operators, on only a *subset* of the entire image sample. If a test fails badly on the sub-sample, it fails by definition on the entire sample. If it succeeds on the entire sub-sample, there is better reason to believe that it should be tested further. Along similar lines, when one analyzes distributions produced for an operator, it is best to test for the simplest, most ideal solutions first, and later for the less ideal partial ones. As the latter require a more complex analysis, and provoke more alternatives to consider, it is best to try to find any easy solutions before descending into the more complicated cases.

A couple of other principles that can be used to reduce the work are to group calculations that need to be shared, and if this is not convenient, to save data that may be needed again to avoid re-calculating it. The first principle is used in the concept of the band operators, which use a single set of image measurements or distributions to analyze four related operations, and in the predicate tests used in proposing operators, which also share some intermediate image results. The second is used on the subproblem node blackboards, where predicate information is stored to avoid re-generating it.

A final way to reduce search is to make use of the mathematical relations between new subproblems and their parent problems, in order to be able to place predicate information directly on the blackboard of a new subproblem, which can significantly speed up the search required to solve it.

All of these techniques help to reduce the overall computational cost of the algorithm search process, without any of them threatening to throw away correct solutions of the simple types examined here. That completes this chapter on some of the theoretical aspects of developing simple set recognition algorithms automatically; in the next chapter I will describe the REM system computer program which was designed to implement and demonstrate some of the ideas presented here.

4

REM System Implementation

The chapter which follows provides a complete description of the REM program, which was implemented to serve as a demonstration of the concepts described in Chapter 3. Not everything discussed in Chapters 2 and 3 has been implemented, but enough has been to serve as a 'proof of principle' of the ideas presented there.

This chapter is divided into five parts. The first discusses how algorithms are represented in the system; the second gives a top-level description of the system code, data structures, and activity; the third part discusses the search control loop; the fourth describes the problem solving mechanism—first giving an overview and then discussing each specific function in greater detail; and finally, the fifth part describes the output produced by the system, particularly that produced by the image displays.

4.1 Representation of Algorithms

Given the fact that the purpose of the REM system was to produce algorithms which solve imaging problems automatically, the first question to ask was how to represent partial and complete image algorithms so that a program could manipulate and think about them. In the REM system, four different levels of algorithm representation are used. These representation levels are structured differently, which makes each one appropriate for a different part of the overall algorithm development process. In brief, we have at one extreme the form in which algorithms must be represented for execution on the hardware. On the other extreme we have the target language which the REM system uses to "think" about solving its problems. In between are two other levels which exist to aid in translating the target level to the executable level.

4.1.1 Machine Level—C4PL Language

To begin, given the nature of the Lisp-Cytocomputer interface that I had available, it was necessary that ultimately all algorithms be *executed* in terms of ERIM's *C4PL*[1] image processing language. This was not entirely optimal, in the sense that C4PL is still an interpreted language, running on the Vax host computer, and there are still two more layers below it before we get to the control software on the microprocessor which actually drives the Cytocomputer hardware. While we have considered developing a closer, more direct interface to the hardware in order to increase speed, what we have available at present is not a direct Lisp link to the hardware, but rather one which talks to the same interactive command interpreter that human users employ in developing algorithms on this machine. One advantage of this arrangement is that it makes it easier to directly compare the efficiency of the system against that of human algorithm developers, since REM displays all of the C4PL code it generates, as it is executed (see Appendix C). An example of a C4PL command sequence to perform a binary erosion residue by a square of radius 5 is given below:

```
MIN3D 222 222 222 5 # X0 -> I1    ; Erosion by Square 5
NOTIMAGE I1 -> I1                 ; Complement
MINIMAGES X0 I1 -> I1             ; Intersect (Min) with
                                  ;  Original gives Residue
```

A longer sequence of small subsequences like that shown above is usually needed to express an entire algorithm.

4.1.2 Lisp Executable Level—MM Language

While C4PL is a very large and powerful language for image algorithm development, what was needed for the purposes of this project at an image command level was a restricted subset of commands which supported just the morphological and other operators that were of specific interest to the project. I also wanted all of these to have a uniform syntax in terms of Lisp calls so that it would be easier to generate executable code in terms of them. I therefore constructed a command oriented (Lisp function) language level, called *MM* (Math Morphology), which has in it all of the basic image processing operations that are, in a sense, the primitives of mathematical morphology from which other operators are derived. Each of these was defined in terms of the appropriate C4PL calls. These 'base operators' are all *uni-directional*, in the same way that a threshold operator which returns only pixels whose value is greater than or equal to a certain constant is uni-directional. (To get those pixels less than the constant requires an additional complement step). The advantages of using uni-directional base

[1]'C4PL' stands for 'Cytocomputer Portable Parallel Picture Processing Language'. C4PL is a trademark of the Environmental Research Institute of Michigan.

operators and then using sequences of additional steps to get their derivative operators are that one has fewer different operators to worry about, all of the base operators and derivatives can be defined and handled in a uniform manner, and it is easier to analyze them mathematically, since most identities are expressed just in terms of the base operators themselves. An example of an MM form for the binary erosion residue by a square of size 5 is shown below:

```
($ERODE X0 SQUARE 5 I1)
($COMPLEMENT I1 I1)
($INTERSECT X0 I1 I1)
```

Again, a longer sequence of such operations is usually needed to represent a complete algorithm. Table 4.1 lists the base operator commands defined in the MM language. MM also contains additional commands, not listed there, which are used to aid in computing coded images during data collection, and for displaying images on the screen.

4.1.3 Algebraic Level—IA Language

The ideal language for representing complete or partial image algorithms, from the standpoint of clarity, is in terms of directed image flow graphs (described in Chapter 3), where images are the nodes and operations are labeled arcs (or vice versa). The advantage here is that one can more easily see the flow of images through the operators, toward the final result. Partial algorithms can be represented by arcs which are not yet specified—they are labeled with subproblem identifiers indicating that they have yet to be solved. Moreover, the effect of modifying a partial algorithm, either by filling in a solution to a subproblem, or by breaking it into two parts, or by solving it partially and then generating a new subproblem—all can be easily understood in terms of physical manipulations and restructurings of the algorithm graph, especially since there are no looping or conditional branching structures with which to contend.

For the purposes of representing partial and complete algorithms in the REM system, I have taken a graph-oriented approach, but in order to represent them symbolically in Lisp, I have developed an isomorphically equivalent algebraic language called *IA* (image algebra), for writing out the morphology code. This language is abstract, and interpretive; while MM and C4PL require expressing each operation as a single command on a single line, with all of the actual image operands specified explicitly, IA makes it possible to write complex expressions on a single line, using variable names as image operands. IA automatically handles the assignment of variable names and intermediate expression results to specific image objects, by maintaining a stack of temporary image registers that it can

```
($ASSIGN set-x set-y)
($COMPLEMENT set-x set-y)

($UNION set-x1 set-x2 set-y)
($INTERSECT set-x1 set-x2 set-y)

($PLUS set-x1 set-x2 set-y)
($MINUS set-x1 set-x2 set-y)

($THRESHOLD set-x val set-y)

($INFMASK-X set-x pos set-y)
($INFMASK-Y set-x pos set-y)

($SUPMASK-X set-x pos set-y)
($SUPMASK-Y set-x pos set-y)

($ERODE set-x shape size set-y)
($DILATE set-x shape size set-y)

($OPEN set-x shape size set-y)
($CLOSE set-x shape size set-y)

($CDILATE set-m set-b set-y)
($CERODE set-m set-b set-y)

($REDUCE set-x connect steps set-y)
($EXPAND set-x connect steps set-y)

($HTHIN set-x connect steps set-y)
($HTHICK set-x connect steps set-y)
```

TABLE 4.1. Base Operator Commands Defined in the MM Language

use for calculations. It also uses infix notation, and is therefore easier to read than the prefix form required by the command-oriented languages. Two ways of writing the same algorithm described above in IA are shown below:

```
   (Y = (X0 INTERSECT ( (X0 ERODE (SQUARE 5)) COMPLEMENT) ))
or
   (X1 = (X0 ERODE (SQUARE 5)) )
   (Y  = (X0 INTERSECT (X1 COMPLEMENT)) )
```

The formal definition of the IA language is given in Figure 4.1.

A real advantage of being able to write algorithms algebraically in this form is that it allows one to concentrate on the essential intent of the calculations rather than on the details of syntax and exactly which parameters need to be used. This becomes even more important when one wants to analyze complex algorithms, to manipulate them, and to build them up in a series of stages. Moreover, many of the important identities of morphology are naturally expressed algebraically, so in a sense, the algebraic form is really the most natural one for representing these kinds of algorithms. Conceptually too, the algebraic form adds to the conciseness and clarity of a written algorithm, not only because it removes some of the inessential details, but because it allows each conceptual unit of the algorithm to be written as a single expression or line of code.

The image algebraic language that I have defined in Lisp has as its primitives the morphological base operations of erosion, dilation, opening, closing, conditional erosion and dilation, etc., along with the operators of union, intersection, clipped addition and subtraction, and complement, which are used to build all of the more complicated derivative sequences that are not primitives. This approach allows IA to reduce the number of different operation primitives to a minimum, which makes it a simple language that is fairly easy to evaluate and to understand. As mentioned above, all of the primitive operators are uni-directional—for example, erosion gives back all of the pixels that are greater than a certain distance from the background, and dilation retains as background all of the pixels which are greater than a certain distance from the foreground. To get the "less than" deriviative sets, additional set combination steps are needed.

4.1.4 Target Level—IC Language

Finally, for the purposes of thinking and reasoning about image operators, it is convenient to group operations derived from the same base operator together. To understand the importance of this, one has to look at a couple of examples. The basic threshold operator gives only pixels greater than

```
IA-Statement == (rsym = ia-expr)

IA-Expr == isym |
        (ia-expr Union ia-expr) |
        (ia-expr Intersect ia-expr) |
        (ia-expr CDilate ia-expr) |
        (ia-expr CErode ia-expr) |
        (ia-expr Complement) |
        (ia-expr E-Epsilon) |
        (ia-expr D-Epsilon) |
        (ia-expr Threshold parm) |
        (ia-expr Infmask-X|Y parm) |
        (ia-expr Supmask-X|Y parm)
        (ia-expr Erode stelt) |
        (ia-expr Dilate stelt) |
        (ia-expr Open stelt) |
        (ia-expr Close stelt) |
        (ia-expr Hthin skelparm) |
        (ia-expr Hthick skelparm) |
        (ia-expr Prune skelparm) |
        (ia-expr Plug skelparm) |
        (ia-expr Reduce skelparm) |
        (ia-expr Expand skelparm)

Stelt == (shape parm)
Skelparm == (connect parm)

Connect == Square | Diamond
Shape == Pdisk | Octagon | Square | Diamond |
         Line-EW | Line-NS | Line-NWSE | Line-NESW |
         Vect-N | Vect-NE | Vect-E | Vect-SE |
         Vect-S | Vect-SW | Vect-W | Vect-NW

Isym == Rsym | Ssym | Tsym
Rsym == X1 | X2 | X3 | X4 | X5 | X6 | X7 | X8 | X9 |
        X10 | X11 | X12 | X13 | X14 | X15 | X16 | Y
Ssym == X0_1 | X0_2 | X0_3 | X0_4 | X_1 | X_2 | X_3 | X_4 |
        A0_1 | A0_2 | A0_3 | A0_4 | A_1 | A_2 | A_3 | A_4 |
        R0_1 | R0_2 | R0_3 | R0_4 | R_1 | R_2 | R_3 | R_4
Tsym == T_1 | T_2 | T_3 | T_4 | T_5 | T_6 | T_7 | T_8 | T_9
```

FIGURE 4.1. Formal Definition of the IA Language

or equal to a value, but those derived from it can give the pixels that are less than some value, or between two values, or outside a range of values. As another example, an erosion by a square of size 5 gives all pixels such that the center of a square of size 5 fits entirely in the foreground. The residue of this gives the locations in the foreground where this condition fails. One can also define band pass algorithms which find places where a size 3 square would fit, but not size 5, and so on.

The point here is that all of these different possibilities which are based on the same operator (or operator/structuring element combination), can be analyzed simultaneously using the same distribution (which is the image histogram in the case of the threshold operation). I will go into the reasons for this in a moment, but what is important here is that for choosing operators to apply in order to solve a problem, it would be ludicrous to treat related operators separately, and thus have to repeat costly image calculations in order to re-create the same distribution several times for analyzing each one. For this reason, I have created a fourth or target language level for algorithm representation called *IC* (Image Concept Language), which essentially groups the related operators together by defining 'band' operators for each base operator. An example of a statement in IC for the binary erosion residue illustrated before is given below:

```
(Y = (XO ERODE-BAND-SQUARE NIL 5) )
```

There is one band operator for each morphological operator/structuring element combination—e. g., erosion by a disk, erosion by a square, horizontal line, vertical line, etc. This is because each of these requires a different distribution to be computed in order to analyze it as a potential solution. One thing this means is that IC is a very large language since it has so many different operators, as opposed to IA, in which the number of different operators has been kept to a minimum, and the structuring elements are factored out of the primitives. Appendix D lists the operators currently defined in the IC language.

Each of the band operators of IC takes an input image and two parameters (one of which may be open-ended) which specify the range of pixels to be selected in terms of the base operation's effects. There are four possibilities for the ranges, as illustrated below:

```
(X OPNAME P1 Nil) -- Right open ended range
(X OPNAME Nil P2) -- Left open ended range
(X OPNAME P1 P2)  -- Inclusive range, P1 to P2 (P1 <= P2)
(X OPNAME P2 P1)  -- Exclusive range, P1+1 to P2-1 (P1 <= P2)
```

Each range requires a slightly different definition in terms of the sequence of IA operators needed to express it, and these definitions vary depending on the class of the operator. Appendix B summarizes these definitions for several different classes—the information contained in this appendix is one of the major contributions of this book, since some of the band operators listed here have never before been formally defined in the literature.

The band operators are the fundamental conceptual units or target language that REM uses to "think" about solving problems. When it proposes possible operations as solutions to a sub-problem, what it proposes are the band operations. Having this language level makes the system more efficient in its analysis, and economizes the code needed to express algorithm ideas, since some of the band operators are fairly complex algebraic expressions. For example, consider the following definitions of two specific IC expressions in terms of the IA language, in their Lisp format:

```
(XO ERODE-BAND-SQUARE 5 3) =
( (XO ERODE (SQUARE 5)) UNION
  (XO INTERSECT ((XO ERODE (SQUARE 3)) COMPLEMENT)) )
```

and:

```
(XO ERODE-CDILATE-BAND-PDISK 6 2)=
( ((XO ERODE (DISK 6)) CDILATE XO) UNION
  (XO INTERSECT
   (((XO ERODE (DISK 3)) CDILATE XO) COMPLEMENT) ) )
```

To come up with the IA sequences, operator by operator, would take much more time for the same result than would working directly with the more concise "package" expressed as a single operator in IC. In short, the IC language represents what I think is the most appropriate level of abstraction for thinking about image operators, in terms of how they might be proposed and analyzed as possible solutions to algorithm subproblems. It is a powerful language for expressing individual steps of an algorithm, while IA is better for expressing entire algorithms in a uniform manner using a minimal number of primitives.

4.1.5 Operation Database

The IA language is purposely small; it has only a few operators to be concerned with, and this makes it easy to translate it into executable form in terms of the MM language. IC, however, is a very large language, as seen from the tables in Appendix D. Luckily the IC operations fall into several distinct classes, such that all the operations in one class can be handled in the same way in terms of the definitions of their band operators, and in terms of how data is collected for them. To reduce the amount of code needed to collect data on images about certain operators and to translate IC expressions into their IA equivalents, REM maintains a small operator database, which has two parts.

First, it defines a hierarchy of important classes of operations, such as 'extensive', 'anti-extensive', 'binary-extensive', 'grey-extensive', 'binary-extensive-pixel-ops', and so on (see Figure 4.2). These classes are used to invoke the definitions required when translating an IC expression. Second, for each IC band operation, certain other information for translating it is also kept on an association list. This includes the IA operator which serves as the basis for its definition, and the structuring element shape which it uses. There is also an entry for the coding function needed to collect data for the operation, which is used by the data collection portion of the system when it wants to generate distributions for each operator proposed by the operator expert system.

4.1.6 Algorithm Representation Summary

To recap, then, the REM system uses four language levels for expressing algorithms—C4PL for actual execution on the Cytocomputer; MM as a Lisp-function command-oriented subset which supports the morphological base operators, and in a sense, is the language for writing executable image algorithms from the Lisp system point of view; IA, which is an interpreted, algebraic, expression-oriented infix-notation language for expressing complete algorithms abstractly before instantiating and executing them with actual images; and finally IC, which is also algebraic, but which groups together related operators under the concept of a band operation, so that they can be evaluated for a solution as a single entity. IC is also used to represent partial algorithm states in addition to complete algorithms. Along with the language definitions, there is a database of operators used to simplify manipulations of IC expressions, and there are translators between levels, so that IC expressions can be converted to their IA (algebraic) definitions, and so that IA algorithms can be instantiated with real image registers, and executed in terms of the Lisp function calls in MM to C4PL. These languages, which define and execute algorithms, form the basis for the rest of the REM system, whose job is to manipulate, construct, and evaluate them. This activity will be described next.

4.2 Overview of System Design

In this section I will present an overview of the REM program by discussing first the major parts of the program code, then the global data structures used by the program, and finally the activity of the system when viewed at the top level. Subsequent sections will present in more detail the lower level activities of the system and the data objects they use.

```
*TRIVIAL-OPS*

*SET-COMBINATION-OPS*

*GREY-TO-BINARY-OPS*


*ANTI-EXTENSIVE-OPS*

   *BINARY-ANTI-EXTENSIVE-MASK-OPS*
   *BINARY-ANTI-EXTENSIVE-PIX-OPS*
   *BINARY-ANTI-EXTENSIVE-CC-OPS*

      *BINARY-ANTI-EXTENSIVE-CC-SKEL-OPS*
      *BINARY-ANTI-EXTENSIVE-CC-PRUNE-OPS*
      *BINARY-ANTI-EXTENSIVE-CC-RES-OPS*

   *GREY-ANTI-EXTENSIVE-MASK-OPS*
   *GREY-ANTI-EXTENSIVE-OPS*


*EXTENSIVE-OPS*

   *BINARY-EXTENSIVE-MASK-OPS*
   *BINARY-EXTENSIVE-PIX-OPS*
   *BINARY-EXTENSIVE-CC-OPS*

      *BINARY-EXTENSIVE-CC-SKEL-OPS*
      *BINARY-EXTENSIVE-CC-PRUNE-OPS*
      *BINARY-EXTENSIVE-CC-RES-OPS*

   *GREY-EXTENSIVE-MASK-OPS*
   *GREY-EXTENSIVE-PIX-OPS*
```

FIGURE 4.2. Operator Class Hierarchy

4.2.1 Program Components

The REM system, which comprises roughly 150 pages of partially commented Lisp code, is basically divided into three major parts. The first is the top level or search control division, which contains the system loader (SYSTEM), the global variable and data structure definitions (SYSVARS), and the system initialization routines (part of REM-TOP-LEVEL). Initialization includes setting up structures both on the Lisp side and in the image processing software. The top level division also includes program sections for displaying, loading, and evaluating the problem to be solved when it is introduced, for carrying out the overall search, and for terminating this search and displaying any solutions found by executing them on the original images, at the end (all part of REM-TOP-LEVEL).

The second division of the system is concerned with problem solving and analysis. This contains the code for calculating the images and masks which define a subproblem, for testing and proposing operations before analyzing them (EXPERT), and for collecting and storing data on the proposed operators (COLLECT). For the collection process, there is an auxiliary section of code (CODING) which, for each class of operator, contains the procedure required to produce a coded image for that operator—one which assigns a class to each pixel based on the operator's effect.

Another part of the second division of the system has all of the functions needed for analyzing the collected distributions for different types of partial and complete solutions (ANALYZE). This section also makes use of an auxiliary library of routines (DISTOPS) for generating, manipulating, and testing predicates on one-dimensional grey and binary distributions. The analysis section of the code is also responsible for generating possible solutions and the program transformations or algorithm modifications that these entail. The DISPATCH portion of this division of the code is responsible for choosing between different proposed solutions and making the appropriate global structure modifications for those selected.

The final major division of the code contains the language definitions and translators for the IC, IA, and MM language levels, along with the operation database (DBASE) which aids in translating IC algorithms into algebraic IA code, and in choosing the appropriate coding function for an operator during the data collection phase.

4.2.2 Data Structures

The global data structures for the REM system, like the program code itself, are divided into three broad groups—those concerned with the overall algorithm search, those concerned with problem solving, and those concerned with image processing or the execution of algorithms. Besides some relatively trivial variables which keep track of the names of files for the current top-level problem that REM is solving, there are five main data structures that are associated with the algorithm search. The most important two of these are the Algorithm Search Tree (AST) and the Sub-Problem Tree (SPT).

Since new algorithm states or graphs are derived by transforming old ones, the set of algorithm states examined by the program forms a tree, with an initial one-arc state at the root. This 'Algorithm State Tree' has an algorithm state description at every node, along with additional information about how well it performs, its associated costs, and the subproblem and image IDs it contains. In a similar way each subproblem the program examines is derived from a previous larger problem, and thus these too form a tree, the 'Sub-Problem Tree', which is used to keep track of problems created and solutions examined. Each node of this tree contains a description of a subproblem, with the original problem to be solved represented by the root node.

The Algorithm State Tree and the Sub-Problem Tree are separate entities for the sake of efficiency, and because activity on them is not always correlated. First, the same subproblem may appear on more than one algorithm state, and the same algorithm state may contain many subproblems; this separation lets us describe each only once. Second, while the most common situation is to have a partially solved subproblem which leads to both a new algorithm state node, and a new subproblem node (representing the part of the problem that remains to be solved), there are other situations where this formula does not hold true. If we fill in a subproblem node with a complete solution for it, a new algorithm state is created but nothing happens to the tree of subproblems; at other times a problem may be split into two subproblems which have to be treated as a connected pair in the Sub-Problem Tree, even though only one new branch is created on the Algorithm State Tree.

At the beginning of the search, the root algorithm state is: (Y = (X0 PROB0)), which simply means that the whole problem remains to be solved. (Ideally Y is going to equal the original accept set for each example image instance if we have found a perfect solution). X0 represents the original image, and PROB0 is the root subproblem, representing the entire original problem.

Nodes on the Algorithm State Tree contain several slots (Figure 4.3). First there is the algorithm state itself, defined in IC, augmented with subproblem identifiers substituting for unknown operators. There are also parent/children connections, lists of subproblem and image identifiers used in the algorithm definition, and slots for holding evaluation information for the algorithm state (accuracy, consistency, and costs).

The Sub-Problem Tree nodes contain a description of the subproblem,

AST NODE SLOTS:	SPT NODE SLOTS:
Parent Algorithm State	Parent Subproblem
P-Nodes	A-Nodes
Images In	X Algorithm
Images Out	A Algorithm
Image Counter	R Algorithm
Algorithm State	Accuracy
	Consistency
Accuracy	
Consistency	Operator Distrib. Data
Costs	Blackboard
Children Algorithm States	Children Subproblems

FIGURE 4.3. Structure of Algorithm Search & Sub-Problem Tree Nodes

parent/child links, the algorithm state nodes which contain that subproblem, the input and output image IDs for the subproblem and a Blackboard for inserting predicate information about that subproblem. Instead of storing the actual images and masks for each subproblem as its description, (which would take up too much space), these images and masks are expressed in the IA language in terms of how to calculate them from the base images and masks of the original problem.

In the course of evaluating operators as possible solutions for the subproblem, distributions for the A and R sets for the given operators are also attached, as is evaluation data for them. Some of this information may be removed when choices among the alternatives have been made—at that point, summary description and evalution data is saved on those alternatives examined, separated into those which were actually chosen versus those that were rejected. This information can be used for backtracking.

The other three data structures involved in the top-level search process are lists—the Agenda, which is a prioritized list of subproblems waiting to be solved (leaf nodes on the SPT); the NewAlgs list, which holds any new algorithm states (leaf nodes on the AST) which were generated by the previous problem solving cycle; and the Solutions list, which holds terminal nodes from the AST which constitute solutions to the overall problem.

Turning next to the problem solving level, there are three lists, one node variable, and one parameter which are of central importance. First of all, the Pnode variable identifies the node on the SPT which represents the current subproblem the system is trying to solve. The Proposals list is a list of IC operators which the expert system on selecting possible operators chooses as ones which should be examined further. The Candidates list contains possible solutions generated by the analysis phase. Each candidate includes new algorithm code and a description of any new subproblems generated as part of the solution, an evaluation of the solution's accuracy and consistency, and a list of any predicates known in advance about the subproblems which could help in solving them. Finally, the Choices list contains those candidates, which, when compared against the others coming out of the analysis phase, were considered to have the most to offer as solutions, for reasons of completeness, accuracy, and/or costs. This final set is the group that will actually be 'dispatched', which results in updates being made to the search level data structures.

The one global parameter used by the problem solver, called *Tolerance*, is a number from 0 to 1000, representing a degree of error that the problem solver will allow in deciding if an operator or sequence of them solves a particular subproblem. Zero means that no error will be tolerated; 1000 means that anything will pass as a solution. This parameter is entered when the user enters the name of the problem to be solved by the system. It allows the user to express the level of 'fudginess' he or she will allow on the part of the system. It reflects the practical fact that solutions to imaging problems are almost never 'pixel-perfect', either because the objects to be recognized have some variability, or because the specification of the desired and undesired pixel masks was not perfectly accurate, or both.

Finally, at the image processing level of the REM system, there are a number of global data structures, used primarily to identify pre-defined images and displays used by the system. First, there are the images which constitute the description of the original problem. There are four instances each, and for each instance there is a base image ($X0_i$), an accept mask ($A0_i$), and a reject mask ($R0_i$), for a total of twelve. The names of these are stored on special array lists to make them easy to access in a loop. Likewise, for the description of the *current* subproblem, REM has a similar group of twelve images, four instances with a base image (X_i), accept mask (A_i), and reject mask (R_i) for each. All of these images are 64x64 pixels.[2]

The image displays are also treated as global data objects (see Section 4.5 for a description of them). There is a set of four windows or images used specifically for displaying problems and subproblems in color with the accept and reject masks used as overlays. There is also a set of nine temporarary image windows which are used for intermediate calculations in the course of executing algorithms on images. The names of these are kept on a stack so that the IA interpreter and other portions of the code

[2]In point of fact, they are particular windows into a larger 480x512 display.

which need them can do so without having the calculations step on each other. The contents of all of these image windows are displayed while REM is executing so that the user can see exactly what the program is doing on the image processing side.

4.2.3 Activity

The activity of the REM system at the very highest level consists of an initialization phase, followed by a simple loop which continues to ask the user for the name of a new problem to solve, until he or she types 'Exit'. Figure 4.4 illustrates the overall flow of control for the top level of the REM system, as well as for the search and problem solving cycles, which will be described in subsequent sections. The initialization phase involves loading specific image processing code into the Cytocomputer image processor, and setting up the look-up tables, pre-defined images, windows, and displays described above, which the system will use in performing its image processing tasks.

When this is done, the program asks the user for the name of a problem to be solved, and a tolerance value to be used while attempting to solve it. Problems are stored as a set of four 64x64 image files, all of which have the same initial name string. Each image is eight bits deep, but the top two of these are treated as the accept and reject masks (A0 and R0, bits 7 and 6, respectively), while the bottom six bits make up the original grey or binary image (X0). These images represent the initial overall problem state.

Once the system has the name of a valid problem, it begins a problem initialization phase, consisting of four steps. First, it initializes the main global variables—the Algorithm Search Tree, the Sub-Problem Tree, the Agenda, the NewAlgs list, the Solutions list, and the Pnode variable. It also initializes the displays on the image processor. Then, it displays the problem to the user in color, by loading the four eight-bit original images into the 'Matte' display and using different look-up tables with the accept and reject masks to really illustrate the task to be accomplished (see Sections 4.5 and 5.3).

When the display sequence is is done, REM loads the problem into the problem solving or 'Overview' display, which involves splitting each original image into its component parts—the base image, X0, the accept mask, A0, and the reject mask, R0. These are then illustrated in black and white. Finally, an initial evaluation is done on the problem, by making measurements on the extent to which each X0 image differs from the desired A0 mask, or contains pixels from the undesired R0 mask. This information is summarized in terms of accuracy and consistency, and is placed on the top level AST and SPT nodes, as well as displayed in graphic form in the Overview display. (It will be used later in the process of proposing solutions

```
Initialize REM

While New Problem Name:

   Initialize Globals
   Display Problem
   Load Problem into Problem Solver
   Initial Problem Evaluation

   Repeat- - - - - - - -
   |
      Prioritize Agenda

      Expand Subproblems Selected:

         Calculate Subproblem Images
         Evaluate Situation
         Propose Operators to Test
         Collect Data on Proposals
         Analyze Data for Range-based Solutions
         Choose Best Alternatives
         Dispatch Choices and Update Trees

      Evaluate New Algorithm States

   |
   Until Done - - - - - -

   Terminate Problem and Illustrate Solutions, if Found

 Stop.
```

FIGURE 4.4. The REM System Overall Flow of Control

to the problem).

At this point, the problem initialization phase ends, and the search phase begins, which will be discussed in the next section. The search phase continues until the system finds a satisfactory solution or gives up (has no more subproblems to solve), and at this point control transfers back to the top level, which will report the result of the search. If solutions were found, it will print them out in IC and IA form, execute them on the original base images, and then overlay the accept and reject masks for each one to demonstrate that the solutions offered do in fact solve the problem. After this, the system actually loads in a *fifth* original image (one it has not seen before), and the proposed solutions are executed on this one as well—this forces the system to try its solution on an example of the same type as the others, but which was not available for analysis as part of the four-element training set.

If on the other hand no solution was found, the system reports this fact and "apologizes" to the user. Whether or not the search was successful, REM loops back to the ask the user for a new example problem to solve, and continues as before unless the user decides to exit.

4.3 Search Management

The primary activity of the REM system consists of trying to solve subproblems and assemble them into complete solutions. The overall search process consists of four steps. First, it has to prioritize the subproblems remaining to be solved on the system Agenda. Next, it attempts to solve some of these subproblems and then evaluates any new algorithm states which were generated from this activity. (Some of these could be solutions to the original problem). Finally, it tests for termination, which may occur if a satisfactory solution has been found, or if it has no more ideas to explore (i. e., subproblems to solve), or if it has run out of time. In this event, it terminates the search, passing control back to the top level which will display any solutions that were found. If termination is not indicated, it continues with the four-step search loop until one of the termination conditions is met.

At the beginning of a new problem cycle, the Agenda contains just one subproblem node, the root node of the SPT, called PROB0, which represents the entire problem to be solved. In this case there is no prioritization to do. As the system attempts to solve problems, however, it may have to propose partial solutions with further subproblems to solve. These are tacked on to the end of the Agenda; they represent leaf nodes in the current Sub-Problem Tree. The system has to decide which of these to pursue first, and which to postpone.

In the current implementation of the REM system, this prioritization step functions as a simple queue, because most of the problems REM currently solves have perfect one-step solutions, or ones which require only a few steps. In a more elaborate system which had to solve more complicated problems, the prioritization step would have to consider several factors in deciding which subproblem node or nodes should be expanded next. As the role of this step is to try and reduce the overall amount of time it takes to solve the top-level problem, it should give priority to those subproblems which are key to the overall solution—either because they are the last step needed for a solution, or because they represent a larger part of the problem (involve more pixels), or because they are part of several algorithm states which have been evaluated as showing good potential for reaching a good (accurate, consistent, low-cost) solution. Additional weight would be given to those subproblems which might be easier to solve because something was known in advance about them (i. e., if some predicate information existed on their blackboards which could reduce the number of alternatives to look at for them).

When the prioritization step is finished, the program begins the problem solving phase, which involves expanding the one or more subproblem nodes having the highest priority, and trying to find complete or partial solutions to them. The specific steps of the problem solving phase are described in the next section. The main points for now are that this phase can cause new algorithm states (AST leaf nodes) and/or new subproblems (SPT leaf nodes) to be generated, as well as the modification of the current SPT node being solved. The new subproblems are added to the Agenda, to be prioritized in the next cycle, but before this happens, the new algorithm states are placed onto the NewAlgs list, and evaluated.

The algorithm state evaluation process has two purposes. First, it has to check to see if any of the new states are terminal states (i. e., complete solutions to the overall problem). This occurs if the algorithm state has no unsolved subproblems remaining on it. These states, (or their AST node names) are placed on the Solutions list, to be used by the top-level termination routine which displays and executes any solutions found at the end.

The second role of the algorithm evaluation process is to assign a utility to each new algorithm state, which reflects the amount of computation it currently requires, and the extent (accuracy and consistency) to which it solves the overall problem. If possible, this value should also reflect an estimate of the amount of work that would be needed to complete this algorithm state, the number of additional algorithm steps which would be required, and the expected final accuracy of the solution. Ideally, the utility assigned to each algorithm state is an estimate of the work, total algorithm cost, and acceptability of the 'best' terminal algorithm state which has the current one as a predecessor. This information can later be used by the prioritization process in selecting which subproblem nodes to expand, as described above.

The algorithm state evaluation process currently used by REM includes only some of the features described above—to begin with, the kind of heuristic information that would be needed to make estimates of the unknowns mentioned above, for this domain, doesn't exist. The current version does check if a new algorithm state is a terminal one, and if so, puts it on the list of Solutions; it also makes an evaluation of the algorithm state, by replacing any unsolved subproblems with the identity operation, and comparing the result produced against that desired. This evaluation information is then added to the new algorithm state node. An evaluation of the costs of the operators has not been implemented at this point.

Finally, when the new algorithm states have been evaluated, the search cycle checks for termination. As described above, the search terminates if a solution has been found, if there are no more subproblems to solve, or possibly if the system has run out of time or even memory space. If not, the search cycle begins again with first the Agenda being prioritized, followed by the expansion of the top priority nodes and the beginning of the problem solving cycle, which will be described next.

4.4 Problem Solving

The process of expanding or trying to solve a subproblem consists of about half a dozen steps (see Figure 4.4). First, the images and masks that make up the subproblem must be calculated from the original images and masks according to the description found in its Sub-Problem Tree node. Next, an initial evaluation is done to assess how far away the problem is from being solved. The distance measure used here is a combination of the percentage of "on" pixels in each image which should be "off", and the percentage of "off" pixels that should be "on". (Grey level pixels, if any, are considered as partially on and partially off). An average is used to combine results from all images in the example set, and a consistency measure based on the variability of the individual results is also returned.

Once the initial evaluation is complete, the operator expert of the system takes over, running a number of predicate tests on the images which make up the subproblem, to determine what relationships hold between the base images and the accept and reject masks for each instance. The purpose of conducting these tests is to determine which operators should be considered as possible solutions to the subproblem; the goal is to keep the number of these as small as possible, because once an operator is proposed for consideration, a fairly complex set of image calculations has to be performed in order to test and evaluate it.

When a set of operators has been chosen to test, the system has to collect data for each of them in terms of what their effect would be on the base images of the current subproblem. The goal here is to get information to decide whether the accept and reject pixels for that subproblem can be discriminated on the basis of a given operator. As a simple example, suppose we are working with grey level images and the operator we want to examine is a simple threshold operation. The data we would need to collect would be the histograms of the accept set pixels and the reject set pixels. If the two histograms have disjoint ranges, we know that we can choose a threshold which completely discriminates the accept set from the reject set. If not, we still may be able to obtain a good partial solution if the histograms do not overlap too much. In a similar way, all of the data collection activities performed by REM return distributions for the accept sets and the reject sets, based on the effect of a given operator and its parameters on the base images of the subproblem. These distributions contain all the information needed to decide whether a given band operator can be used to solve a sub-problem, and to what extent.

When distributions have been collected for all of the proposed operators on all of the subproblem images, the problem solver begins its analysis and selection phase, in which it analyzes the distributions to see which of the operators offer the best solution to the subproblem. Ideally, this activity is performed in a series of stages or 'rounds' in which the system first tries simple tests for complete solutions before going on to more complex tests for partial solutions. For example, the system could first look for perfect solutions, and if it found any it could stop right there. If not, it would go through a second round, to look for close-to-perfect solutions. Next, it would check to see if any operators offered a partial solution that still maintained an inclusion relationship (Sup or Inf) between the actual result and the desired result. In a fourth round, it would drop this requirement, and so on. Eventually it would come up with a list of possible alternatives (Candidates) and would have to decide which one or more of them offered the most promise for the least cost. These would then be entered on the Choices list, to be dispatched, which would enter them as official solutions to try.

Currently the analysis phase looks for perfect solutions to a subproblem, or for certain directives issued by the operator expert as to how to transform the problem, so as to put it into a form that will be easier to solve. For example, the operator expert may recognize that the problem is already solved, or that a complement operation is needed to either solve the problem or to put it into a preferable form, or that the problem can be solved perfectly if certain noise points are ignored—in these cases, data collection is usually skipped, and it issues a directive for the analyzer to simply follow, because the latter is ultimately responsible for all code generation and algorithm transformations. A consequence of this is that the selection or choice phase currently has little to do other than pass on any perfect solutions or directives which were generated by the analyzer.

Current efforts on the next phase of the implementation include having the analyzer also look for solutions which are not quite perfect, but for which the amount of overlap in the ranges of the distributions is within the tolerance specified by the user. Another fairly straightforward case is when the analyzer finds that two distributions are disjoint, but not over a single range—this requires constructing a partial solution over one range, then creating a subproblem to apply the same technique one or two more times over different ranges and combine the results to solve the problem completely. A third case being worked on at present is one in which the results of two commutative operators (such as the X and Y directional infmask and supmask operations) can be applied at the same time (creating an image sub-window) when it is determined that both can help restrict the solution. This would avoid creating a separate subproblem and search cycle to find the second one of them.

When these efforts have been completed, the analyzer will be set up to look for true partial solutions, where the accept and reject sets can only be discriminated to some large but incomplete degree, and the creation of a subproblem to solve the remaining part is necessitated. At this stage, the role of the final selection mechanism will become more complicated and crucial, as several different partial solutions may be offered as candidates to a given subproblem, and it will be the job of the selection mechanism to pick the best ones, based on considerations of solution completeness (how many subproblems were generated), as well as accuracy, consistency, and eventually the costs of the operations utilized.

Once the choices are completed, the current subproblem node is essentially finished, and can be cleaned up and closed out. Distribution data and facts which are no longer needed can be removed, leaving only summary information about the solutions proposed and the results obtained, which might be needed for future reference or in case some back-tracking had to be performed later. (For example, if no excellent solutions to the original problem could be found, the system might be set up to relax its criteria and try again).

The dispatching process consists of generating new Algorithm State Tree nodes by substituting new code for the problem being solved (including any new subproblem arcs), generating new subproblem nodes for the Sub-Problem Tree if any were created, putting the new algorithm states (AST leaf nodes) on the NewAlgs list for overall evaluation, and putting the new subproblems (SPT leaf nodes) on the Agenda to be prioritized and eventually solved. The updating of the two trees is somewhat complicated by the fact that cross-references have to be maintained between them, so that each algorithm node knows the names of all the remaining subproblems it contains, and each subproblem node knows all of the algorithm states of which it is a part. If no solutions were found by the analyzer, or if all of them were rejected by the selector, then no updates are made by the dispatcher and the current subproblem is closed out as being unsolved. This could potentially lead to the removal of those AST nodes which contained them as also being unsolvable, unless the possibility of backtracking or later analysis would benefit by simply making them dormant or of low utility temporarily.

This then completes the problem solving cycle. The following subsections elaborate on what has been described here conceptually, by going into more detail about the different problem solving components of the REM system and their present state of implementation.

4.4.1 Subproblem Instantiation

The first step in the problem cycle is to calculate (or instantiate) the X/A/R image triplets which define the subproblem, and load them into the appropriate image windows to be displayed. The algorithms to do this are stored on the SPT node which represents the subproblem, as IA algorithms. These algorithms are expressed solely in terms of the original images and masks—X0, A0, and R0. Representing the subproblems as algorithms on the original images, instead of directly as the subproblem images themselves, was done to save on storage. At present this is not a real concern, but later on if the system worked on many image instances at a time, and had many unresolved subproblems, storing these images could rapidly eat up available storage space, especially if larger images were used.

As a result, one of the somewhat complicated aspects of the present system concerns the correct updating of the algorithms to compute the subproblem images. For the case of the base image X, what we want is just that part of a total algorithm state graph needed to compute the input image to the link representing the current subproblem. One way to do this would be to have a procedure which could take a state graph, start at the node representing this input image, and work backwards extracting all links and nodes until the original X0 input. Since, in fact, new subproblems are derived from old ones, what really has to be done is to take the algorithm for X0 from the parent subproblem, and modify it based on the program transformation which was made to get to the current subproblem. This is accomplished in the analyzer, described later on in this chapter.

A more difficult problem is posed by the case of the accept and reject masks, A and R. Again we want to modify the algorithms in the parent subproblem in order to get their new definitions for the new subproblem. This often requires using the operator applied in order to get the new base image. For example, suppose we have an A-Sup condition, so we just need to remove certain pixels from the foreground. We may have an operator which can remove some but not all. The new subproblem requires that we find a way to remove the rest. How do we express the new X, A, and R sets in terms of the old? The new X set will be the old one followed by the anti-extensive operator used to eliminate some of the unwanted pixels. The new A set is the same as before—we still are trying to get the same pixels. How do we express the R set? The new R set is just the pixels left that we don't want—specifically, it is the intersection of the old R set, with the old X modified by the operator to get the new X. (Note that we can't define it in terms of the new X, since it ultimately has to be expressed only in terms of the original images and masks).

What this means is that we have to construct an algorithm which is the intersection of the previous R algorithm with the previous X algorithm modified by the latest operator. If we are very deep into the game here, each of these algorithms may have several substeps, and may use the same names for some of the intermediate images. Such conflicts have to be resolved gracefully in the course of constructing new algorithms for A and R. In addition, while I have cited only one example here (A-Sup), there are many other situations which can occur, depending on the exact form of the partial solution used to solve the parent subproblem. For example, the A-Inf, R-Sup, and R-Inf cases all have to be handled differently, and the method of generating the new masks will also vary depending on whether we are doing a sequential decomposition of the current problem or a parallel (i. e., union or intersection based) one. Much of the work I have done in this area has not yet made it into the implementation since the only partial solutions it covers at present are those which require modifying a mask to ignore certain pixels, and complementing the base image to produce an A-Sup or A-Inf condition.

4.4.2 Subproblem Evaluation

In order to evaluate the state of a subproblem, or alternatively, to evaluate how well an operator solves a subproblem, we have to have a reliable measure that we can use. For problems involving sets, the natural measure is one based on the errors (false alarms and misses) that are present. In the first case, we want an idea of how to proceed; in the second, we want to decide whether an operator does a good job. In both of these cases, there are two main criteria to be concerned with—*accuracy* and *consistency*. We want solutions to be as close as possible to the desired result (or its complement), but since any problem is composed of multiple example images, it is also important that the results be somewhat consistent from image to image—that is, an operator should not do very well on some images and very poorly on others, since this suggests either the problem needs to be decomposed into different parts, or that perhaps the operation under consideration does not really capture what is truly going on in the images, except by coincidence in some cases. While consistency is clearly a secondary consideration, it does provide a way of distinguishing between solutions whose average accuracy is similar.

When we evaluate a problem's status or the ability of an operator to solve a sub-problem, the Accuracy for a given image is expressed as a distance from the desired condition, which is to have 100% of the accept set pixels in X, and 0% of the reject set pixels. (The exact opposite is nearly equivalent, since a simple complement operation can then be used to get the desired result). What we normally have is some mixed percentage of the accept and reject pixels in X (the foreground), and some other mix in X-complement (the background). These percentages will also typically vary from image to image. The space that contains all of these possibilities is the unit square. In REM all evaluation measures are scaled to the interval 0 to 1000, so we can think instead of a square between (0,0) and (1000,1000), as shown in Figure 4.5. As I have set up this square, the upper right corner or goal state corresponds to 100% A in X and 0% R, while the lower right corner corresponds to 0% A and 100% R. The top edge of the square corresponds to those situations in which 100% of A is in X (A-Sup), the bottom edge to 0% of A in X (R-Inf), the left edge to 100% R in X (R-Sup), and the right edge to 0% R in X (A-Inf).

Since each problem is made up of several image instances, what we really have is a set of points within the square. To compute the Accuracy of the set with respect to A, we find the average of the distances of each point to the upper right corner representing all of A and none of R. The Accuracy with respect to R is based on the average of the distances to the lower left corner representing all of R and none of A. Accuracy increases as the average distance decreases, and the results are scaled to run between 1000 and -414 or $1000 \times (1 - \sqrt{2})$.

The Consistency is the extent to which all of the points for a given evaluation are clustered together. This is computed by finding first the centroid of all the points in the unit square space, and then finding the standard deviation of the points' distances from this centroid. This value is then scaled to 0 to 1000. The unit evaluation square, along with the points representing the percentages of A and R in X, and the centroid, are displayed in the Overview image display (see next section), when the initial evaluation is done on the problem. The location of the centroid in the unit square is used by the system to decide on which type of overall strategy to pursue, as we will see next. In some cases the initial evaluation can be skipped, if the system knows already what the results will be. For example, if we have an R-Sup situation which we want to convert to an A-Inf by a complement, we know what the result of the new evaluation will be without measuring the images, so these calculations are skipped.

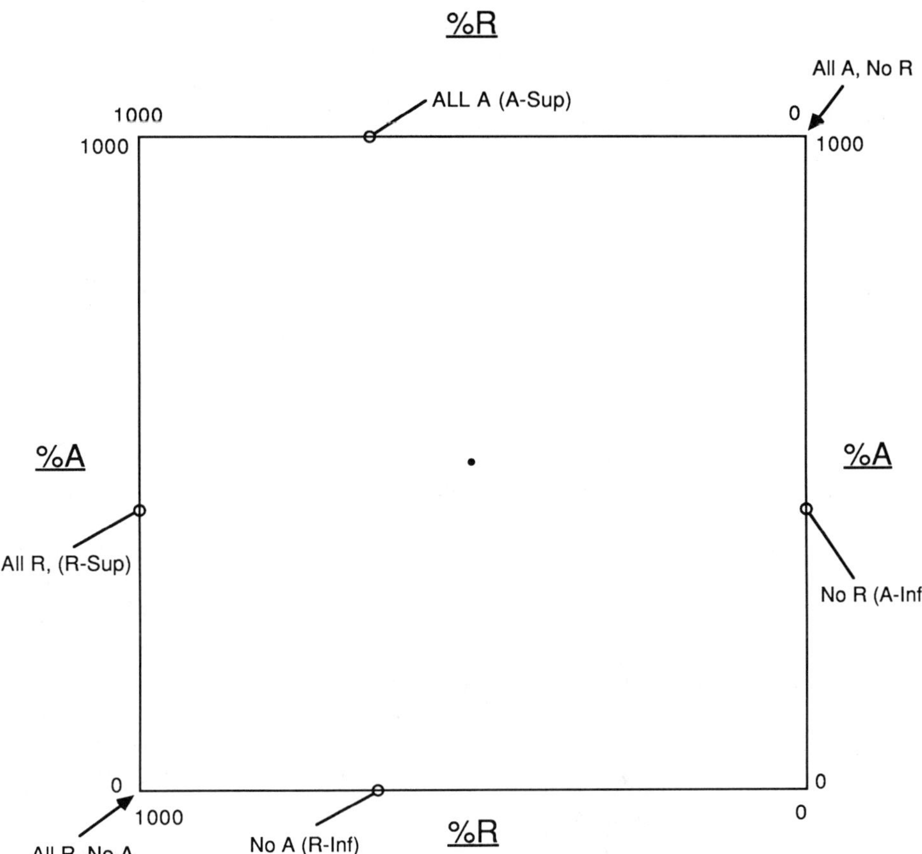

%A is percent (0-1000) of Accept Set pixels in X.

%R is percent (0-1000) of Reject Set pixels in X.

FIGURE 4.5. Unit Square Evaluation Space

In general, if the centroid is near the top edge (close to A-Sup) then the strategy is to use an anti-extensive operator; if it is near the right edge (A-Inf) then an extensive; if near the bottom edge (R-Inf), a complement followed by an anti-extensive; and if near the left edge (R-Sup), a complement followed by an extensive operator. These general strategies hold even if the condition isn't perfectly satisfied, but then it will also be necessary to generate additional subproblems to pick up the remaining parts. As the centroid approaches the middle of the space one is more inclined to split the problem into separate parts, whose solutions would be combined by a later step.

4.4.3 Operator Testing and Proposal

When the initial evaluation is complete, the operator expert, in the form of a decision tree along with a large number of routines for calculating important predicates, takes over control by trying to propose operators which can best solve the problem. The decision hierarchy is shown in outline form in Figure 4.6. The form of this proposal-generating expert is an elaborate branching structure of tests and contingencies, rather than a homogeneous set of if-then rules to be executed independently in parallel. The reason for this is that many of the tests only make sense as refinements of other, previous tests, and because there is a desire to limit the number of image predicate tests that have to be performed to an absolute minimum, since image operations are what take the most time. A non-hierarchical set of rules could require repeating or storing many shared image calculations.

The operator expert begins down the decision tree by asking whether the examples are binary or grey. If they are grey, it will immediately propose the threshold operator to try since at present, that is the only one it has for grey level images. If the images are binary, it looks to see if one of the preferred, strict containment conditions (A-Sup, A-Inf, R-Sup, or R-Inf) holds. This it will find out either by checking the blackboard for the current subproblem, (on its SPT node), or by checking the centroid value calculated in the evaluation step (also on the SPT node in a separate slot).

I should mention at this point that all of the predicates defined in the operator expert first check to see if the answer is already known (true or false) on the subproblem blackboard, before trying to calculate an answer either by going after stored data in other slots, or in the worst case, by doing calculations on images to establish truth or falsity. In many cases the same question may have to be asked more than once; in others, it may be possible to add some predicate knowledge to the blackboard of a new subproblem when it is created, based on what was true for the parent subproblem, and the relationship between the two. In both situations, having the predicate function check the blackboard before calculating saves time. Moreover, any

```
Grey => THRESHOLD

Binary

  A-Eql => IDENTITY
  R-Eql => COMPLEMENT

  A-Sup (Anti-Extensive operations)
     Position ops
        Sup-XAddr => INFXMASK
        Sup-YAddr => INFYMASK
     Particle ops
        Length => REDUCE-CDILATE SQUARE
     Basic MM ops
        Distance => ERODE PDISK
        Size => OPEN PDISK

  R-Inf => COMPLEMENT + Anti-Extensive Subproblem

  A-Inf (Extensive operations)
     Position ops
        Inf-XAddr => SUPXMASK
        Inf-YAddr => SUPYMASK
     Hole ops
        Length => EXPAND-CERODE SQUARE
     Basic MM ops
        Distance => DILATE PDISK
        Size => CLOSE PDISK

  R-Sup => COMPLEMENT + Extensive Subproblem

  OR (Tolerance Tests):
     Almost-A-Sup, Almost-R-Inf
     Almost-A-Inf, Almost-R-Sup
               => REVISE-PROB (Ignore 1-sided noise)
```

FIGURE 4.6. Operator Selection Decision Tree

time a predicate function has to calculate its answer, it places the result on the blackboard to prevent this from having to be done again.

The system checks the containment predicates together, and puts the answers it finds onto the blackboard. (Note that A-Sup and R-Inf are mutually exclusive, as are A-Inf and R-Sup). Then it checks to see if the A-Sup and A-Inf predicates are both true, in which case it proposes the identity operation, or if the R-Sup and R-Inf predicates are both true in which case the complement is the correct answer. If only the A-Sup condition is true it will start to look through the anti-extensive operators, while if the A-Inf condition is true, it will look through the extensive ones. If an R-Inf condition holds, it will immediately propose a complement operation, since this will produce an A-Sup condition, and likewise if an R-Sup condition holds, it will propose a complement to convert this to an A-Inf condition.

If neither of the four strict containment relations holds, the system will use the Tolerance parameter to check and see if one of these four holds within the tolerance specified. If so, then it will propose a Revise-Problem directive to change the A or R masks so as to make one of the strict containment relationships hold. (The directives to revise the problem or to perform a complement and create a new extensive or anti-extensive subproblem are picked up by the analyzer portion of the code).

If the operator expert has decided that an extensive or anti-extensive operator is appropriate, it will next try to narrow the field by testing some additional predicates, now directly on the images. It begins by drawing a subsample of two images from the original set of four, which will be used for these tests.[3] If a necessary condition fails on these two images, it will fail for the entire sample as a whole; if it succeeds on both, it suggests, but does not guarantee, that it might succeed on the sample as a whole. Taking a subsample for testing thus reduces the amount of computation and time needed, during this phase of proposing possible operators.

In testing for particular classes of operators, the expert will first check if one of the position mask operators might be appropriate. Basically it will test the images to see if there is no significant overlap in the X or Y directions between the addresses of the A and R sets. If not then it proposes the appropriate operator as a possible means of separating them. This test does not guarantee that the two can be discriminated in this way, but it is a necessary condition, and thus provides the possibility of a solution. Following this, the operator expert does a number of fairly complex image calculations, to try to establish the answers to four predicates. The image calculations are shared because these predicates all share some of the same

[3]Naturally, drawing subsamples for testing would make more sense if the system were working with an input sample of tens or hundreds of images, but the efficiency principles are still the same.

intermediate results, which would require re-calculation if these predicates were performed independently.

For the anti-extensive case these predicates are as follows (a similar, dual set of predicates exists for the extensive case): The first one checks to see if there were any example images where the individual particles were either in A or R but never mixed between the two. This might suggest a particle operation but it could also be produced by an opening. The second predicate asks if *all* example images showed all particles to be either in A or R—never mixed. If true, this dictates a particle approach. If false, and the first one is true, then an opening would be selected.

The third predicate asks if the borders or inside contours of the particles in each image are partially in the A set and partially in the R set. This is likely to occur with an opening but cannot occur with an erosion by a compact (as opposed to a line-shaped) stelt. Finally, the last predicate asks if the borders are never mixed between the A and R sets, which is reasonable for an erosion but impossible for openings by compact structuring elements.

This completes the analysis of the current operator expert, except for the recent addition of a fifth predicate to this last group. I found in the course of testing that there are cases when a problem which is really based on a position mask will also trigger the third predicate that suggests an opening. The new predicate only allows an opening if also the smallest size-classed pixels were never mixed between the A and R sets—something which is required of openings but unlikely if a position mask is responsible. I am also currently working on ways to select between a diamond, square, and octagon or disk-shaped structuring elements for the basic four morphological operations. Distinguishing between the different particle and hole (connected component) operations, related to length, width, and surface roughness, appears to be a more difficult task.

While proceeding through these predicate tests, the REM system displays all of the image calculations required and prints all of the results and conclusions that it derives from them, as they are produced.

4.4.4 Data Collection

The data collection phase of the program is fairly straightforward. First of all, any directives (Identity, Complement, Revise-Problem) cause the data collection to be skipped since it is not needed. In all other cases, there are four steps to perform for all of the IC operators proposed by the operator expert. First, the coding function associated with the operator has to be run over each image to produce a coded image. Examples of these for different operations are shown in Figures 4.7 and 4.8, for the original image illustrated in Figure 2.5. A spectrum lookup table (black background, red (low) to purple (high)) has been used to make it easier to see the different levels of these images.

Second, the coded images in their grey form are overlaid with the accept (A) and reject (R) masks. Third, histograms are taken for the parts of the coded images which are under the accept set and the reject set, respectively. Finally, these distributions for the four example images have to be stored on the SPT node of the current subproblem, under a property named by the operator from which they were derived. This will allow them to be retrieved later by the analyzer. All of the information needed to decide if the given operator can successfully discriminate between the A and R sets is contained in these distributions—they represent a significant reduction of the image information, along the dimension specified by the underlying operator.

The system displays all of the coded images as it executes them, as well as the overlays of the accept and reject sets, and the distributions collected for them (see Section 5.3). In particular, the overlay of the accept sets in green and the reject sets in red, transparently on the grey level coded images, gives a visual idea of whether the two sets can be separated on the basis of a range of values from the coding or classification of pixels by the given band operator, or not. This result is then confirmed by looking at the distributions, which are displayed subsequently.

The coding functions produce many different types of classification of the image pixels, as is illustrated by Figures 4.7 and 4.8. The first two panels in Figure 4.7 show the coding functions produced on a binary image shown earlier for the anti-extensive X and Y position mask operations. Below this are the ones for the erosion by a disk and a square, and below that for the dilation by a disk and a square. (Black and white versions of the erosion and dilation cases, along with the associated distributions, were shown previously in Figure 2.7). In Figure 4.8 we see at the top the coded images for openings by octagons and squares, followed by closings by the same structuring elements. Note how much these differ from the erosion and dilation. (Black and white versions of the opening and closing coded images, along with the associated distributions, were illustrated in Figure 2.11). The last two panels illustrate coded images for operators which classify entire particles, in the first case by their maximum width and in the second by their maximum interior length.

4.4.5 Data Analysis and Solution Generation

Following the data collection phase, the analyzer takes over. The analyzer is one of the most important parts of the entire system, as it has to perform four major tasks. First, for each band operator proposed by the operator expert, it has to analyze the distributions collected in the previous step, to see if there is a way to completely or partially distinguish the accept and

FIGURE 4.7. Coded Images for Position Mask, Erosion and Dilation

FIGURE 4.8. Coded Images for Opening, Closing, and Particle Ops

reject set pixels, over all example images, by selecting a range of parameter values for that band operator. If there is, it has to evaluate how well the operator performs this discrimination, and then it has to generate the code, or program transformations, necessary to replace the subproblem link in an algorithm state graph, with a solution based on the given operator.

In addition to generating the code and the summary evaluation data, it also has to set up any new subproblems mandated by the form of the proposed solution, including new algorithms for calculating the X/A/R images which define the subproblems, and anything that is already known about them such as predicates they satisfy or do not satisfy, or operators to consider or avoid in solving them. All of this information is then packaged up into a structure called a 'candidate' solution, and put onto the Candidates list to be selected from in the next stage.

More specifically, a candidate is composed of the code to be substituted for the current subproblem, including any new subproblems, an evaluation data summary (accuracy and consistency) of the ability of the operator to solve the problem, a formal list of new images and of new subproblems generated by the solution, a list of new X/A/R algorithms which define the new subproblems, and a list containing any predicate knowledge that is true about them. The role of the analyst is to prepare this package which contains everything needed to perform the program transform and to update the trees. It doesn't actually do the updates because the packages have to be judged against one another before being selected and dispatched.

In a little more detail, the analyzer first checks the list of proposals coming from the operator expert, for any specific directives such as Identity, Complement, or Revise-Problem. These don't require a real analysis, but simply have to be followed. In the case of Identity, for example, it proposes a candidate package directly. The program transformation just substitutes Identity for the old problem, the evaluation data are all known in advance, and all the other lists are empty. Similarly if Complement is proposed and it is true that the current X is always equal to the reject set, then a canned package can also be prepared and sent out.

If the Complement operation proposal is due to the discovery of an R-Inf or R-Sup condition, more has to be done. In this case the program transformation code includes a complement but it also requires another graph link with a new subproblem. The evaluation data aren't pre-determined but they can easily be calculated based on those of the current problem when it was initially evaluated. One new image identifier and one new subproblem are generated, in formal terms only. The algorithms which define the new subproblem based on the original images have to be generated; specifically, the algorithm for the current X image has to be complemented for what will be the new X image. Finally certain predicates have to be placed on the knowledge list for the new subproblem, based on whether the current problem, as indicated by information on its blackboard, satisfies an R-Inf or R-Sup condition.

In the case of a Revise-Problem directive, the new code will simply replace the old subproblem ID in the algorithm graph with that of the new subproblem, the evaluation data stay the same as for the current subproblem, and no new image IDs are created. What happens in this case is that one of the A or R set algorithms have to be modified in a way that will change the mask, and at the same time change a near A-Sup, (R-Inf, etc.), condition into a strict one. Furthermore, it has to put predicate information about the new conditions which hold and do not hold onto the knowledge list.

This is how REM creates candidate packages in a few very specific cases; in the more usual case, it has to actually perform some data analysis before constructing the candidate solution. First, it computes something called the 'Union of V-grams', which is just the union of the projections of the distributions taken for all four example images. This tells us all of the values taken by any of the pixels in either the A set or the R set, over all of the coded images. If these two are disjoint, which is the first test performed, then there is a possibility that a solution of an appropriate type (i. e., based on a *range* of values) may exist. If they are not disjoint, no complete solution can exist so the operation is rejected immediately.

At this point the analyzer does additional calculations to see if there exist parameters for one of the four types of allowable ranges, which will completely separate the accept set from the reject set. If so, it takes these two parameters, constructs a range from them, and then generates the code for a band operator based on that range. It also conducts an evaluation of this operator as a solution by checking the accuracy and consistency of this step as a solution to the subproblem under consideration. (This evaluation is superfluous when a complete separation exists, but I run it anyway to excercise the code, in anticipation of using it later with partial solutions). In the case of a complete solution, no new images or subproblems are generated so all other sub-lists in the candidate package are empty.

I am currently adding other situations, described briefly above. For example, if a small overlap appears in the distributions, one which is always less than the allowable tolerance, then system will go ahead and construct a solution anyway, instead of giving up as it does now. Or, if it encountered a situation in which a single operator could solve the problem (had disjoint A and R distributions), but not in terms of a single band operator, the analyzer will be able to construct a multi-step solution from repeated applications of the same operator. Following the addition of these cases, the next step would be to add analyzer components which would look for true partial solutions, and generate new subproblems to cover the parts remaining to be solved. An important subcase of this type is when one can pick a range for the associated band operator, such that one of the four preferred containment situations (A-Sup, etc.) holds—in this case we only have to generate one subproblem, and we know which class of operator to apply to it. In a second subcase no good range exists for this purpose, and one is then forced to generate two subproblems, one to add in the missing pixels and another to remove unwanted ones. Another alternative could be to split the current problem into distinct parts along some other dimension, to be recombined later on.

Such extensions would bring up many interesting issues, in terms of how to generate the new algorithm code, how to find the best ranges, how to generate the new X/A/R algorithms, how to decide which strategy to use in creating subproblems, and what knowledge to include on the blackboards of the newly created subproblems. Another extension, which would be useful in the case where a backtracking mechanism was implemented to allow re-examination of solutions, would be to place a note about all operators rejected by the analyzer into a slot on the curent SPT node, telling briefly 'why' each was rejected.

4.4.6 Selection and Dispatching of Solutions

After the analyzer has generated one or more candidate solutions, the next step in the solution of a subproblem is to select between them. In the general case, these may differ in terms of the number of subproblems generated (one or two), in the highest, average, or lowest accuracy, in the consistency, or in the costs of the operators used. A general selection mechanism would have to weigh all of these factors and decide which one or more of the candidates generated offered the best benefits for the costs. Only these packages could be passed on to the dispatcher to be actually instantiated. Summaries of the solutions generated but not selected might be placed in the current subproblem node, much like those rejected by the analyzer described above, in the case of a system which did backtracking. This would avoid re-calculation of the same information, in the event that the system wanted to reconsider a subproblem solution. In the current REM system implementation, there is never more than one candidate solution offered by the analyzer, so there is no comparison to do—the selector simply passes the candidates chosen, if any, to the dispatcher.

In a future implementation of REM, one which included consideration of partial solutions to subproblems, the best strategy for analysis, generation, and selection of possible solutions would probably be to have the analysis and selection process operate in a series of stages or 'rounds'. More perfect solutions are, at the same time, more desireable and usually easier to check (require less computation), than imperfect and partial solutions which require a more complicated analysis and the consideration of many more alternatives. Because of this, the desire for search efficiency dictates that we first look for perfect solutions, then imperfect ones, then partial ones and so on—comparing and selecting at each stage—only checking the worst cases if we are forced to do so. By performing analysis and selection of candidate solutions in a series of rounds, we could avoid the costly examination of many imperfect or partial solutions, in the case where some other proposed operator offered a very good or complete solution. The installation of a backtracking mechanism could also be part of this scheme—certain levels of solutions wouldn't be considered if a satisfactory solution was found at a higher level, *unless* subsequent examination of that solution failed to converge, in which case the system might want to back up and consider some of the other less ideal solutions to the given problem.

Moving on to the dispatcher, the role of this part of the code is to instantiate candidate solutions selected in the previous step, by generating new algorithm state and subproblem nodes, and by making new entries on the Agenda and the NewAlgs list. Specifically, the dispatcher has to first find all of the AST nodes which contain the current subproblem in their graphs. To make this easy, a list of these is maintained on the subproblem node itself. Then, for *each* of the chosen solutions, it has to create a new AST node which instantiates that solution on every one of the AST nodes which contain the current subproblem. Since some of the solutions may require the creation of new subproblems, however, before it can actually generate the new AST node based on a given solution, it has to generate the new subproblem nodes that the solution requires. This involves several steps.

For each subproblem required by the given solution (one or perhaps two), it has to generate a new SPT node object, attach it as a child to the current subproblem, fill in the IA algorithms for calculating the X/A/R image triplets that define the new subproblem, and enter any a priori knowledge about this subproblem on its blackboard. The latter two elements are part of the candidate package; all other slots are left empty at this point. Then the actual names of the new SPT nodes are entered onto the Agenda and an association of these names with their formal names in the Candidate package ('P1' and 'P2') are passed back, so that they can be substituted where they appear in the new algorithm code.

Once the new SPT nodes are generated, the given solution can be instantiated on all of the AST nodes which contain the current subproblem as a link in the graph. First a new AST node object is created, which is attached as a child to its parent AST node. The list of SPT nodes on the new algorithm state is filled in by removing the current one from the list found in the parent's slot. Then new cross-references are established between the SPT and AST trees by adding the new subproblem node IDs to the SPT node list of the new algorithm state, and by putting the name of the new AST node object on the AST nodes slot for those new subproblems.

Now the old algorithm state can be updated with the new IC language code to be substituted. To accomplish this, new actual image IDs have to be generated (X2, X3, etc.) to serve as intermediate results in the new algorithm state. In the Candidate list, these are referred to by formal names (I1, I2). The actual names have to be ones that aren't already being used in the current algorithm state, and they have to be associated with the formal names. In addition, the actual input and output image IDs which begin and end the link in the algorithm graph that contains the current subproblem, have to be associated with the formal names 'In' and 'Out'. In the next step the new code, which was really a formalized abstraction of the code to actually appear in the new algorithm, has to be substituted with actual image IDs and the actual new subproblem IDs. When this is done, the dispatcher fills in slots on the new AST node which keep track of the input and output image IDs for every subproblem remaining on it. Finally, it takes the instantiated new code and uses it to replace the link in the parent algorithm state containing the current subproblem ID. This will then be the algorithm state for the new AST node. Other slots are left empty for the present, and the generation of the new AST node terminates by placing it on the NewAlgs list to be evaluated at the end of the search cycle.

When all of the chosen solutions have been instantiated in this way, on all of the AST nodes which contain the current subproblem, the dispatcher closes out the current subproblem by removing any data on it which is no longer worth keeping, and then it terminates, passing control back to the search manager, whose next task is to evaluate the new algorithms and test for search termination, which were described in Section 4.3.

This then completes the description of the REM system implementation and some of the forseeable extensions to it. Following a brief description of the output produced by the system in the next section, the next chapter will be reviewing its present capabilities, including the results of testing and evaluating it on a large number of example problems, and examining its efficiency compared to humans and to other search techniques.

4.5 System Output

The REM system produces output simultaneously on the terminal screen and on the image processing display screen. On the terminal screen it prints all of the C4PL (image processing) code sequences that it executes (see Appendix C). Most of this code is not hard-wired, but rather has to be interpreted by evaluating Lisp expressions for specific image and size arguments. REM also prints out all of the IA code that it interprets, and the subsequent MM calls which implement this code. In addition, it prints out

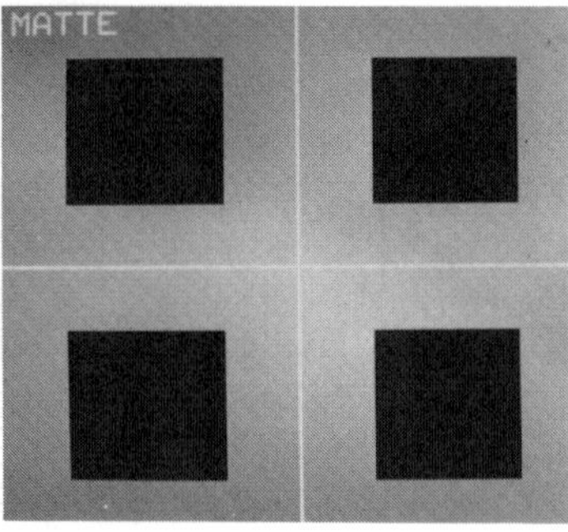

FIGURE 4.9. Matte Display

numerous messages about where it is in the program code, what it is doing, what it finds out and the contents of important data objects. There is also a problem display sequence in which it describes each problem to be solved to the user, and a termination sequence in which it lists and illustrates any solutions that have been found (see Section 5.3).

On the image display side, there are two major displays used. The first, called 'Matte', is a four panel display which has windows for four 2x magnified images (Figure 4.9). This is used to illustrate image subproblems when they are introduced, and to display solutions at the end. The second, larger display is used by the problem solving apparatus to run and illustrate its activities. It is called 'Overview' and has four parts (Figure 4.10.). The upper left quarter has a large window for 4x display of individual images that are of particular interest. On the lower left is a panel of 4 by 3 64x64 image windows, which are used to hold each of the four instances of the original problem images and masks, i. e., X0(1–4), A0(1–4), and R0(1–4). On the upper right is another panel of 12 windows which hold the four instances of the X, A, and R images that define the *current* subproblem being solved. Finally, at the lower right is a panel of 9 windows which are kept on a stack and used for all image calculations. This allows the viewer to see these calculations being carried out, since C4PL makes it possible to have the Overview display automatically updated whenever one of the windows in it is modified. These updates slow down the program somewhat, but they make the activity of the system more understandable and interesting.

All of the windows defined in the two displays are physical image objects defined in C4PL and kept on special lists by the REM system in Lisp. These are displayed on a graphics device having an 8-bit pixel range with three 8-bit look-up tables, one for each RGB color. As mentioned earlier, the system uses 64 by 64 images exclusively, which are either grey level or binary. Six bits are used to hold the images; the two other bits in each

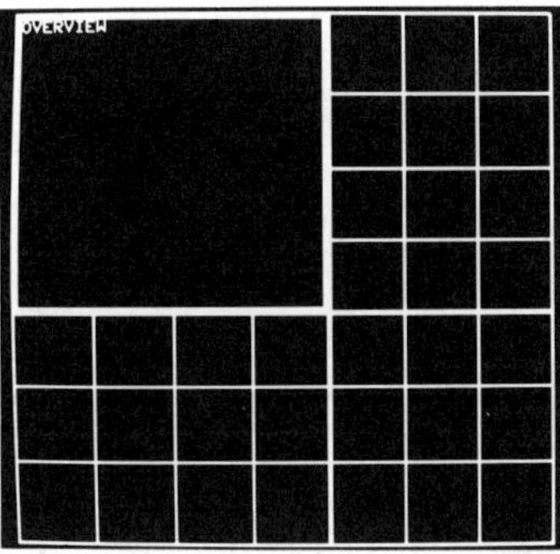

FIGURE 4.10. Overview Display

pixel are used for bit plane overlays—bit 6 for the R set and bit 7 for the A set. Having 6-bit images with two more upper bits for overlays effectively divides the 0–255 dynamic range into 4 parts: 0–63, 64–127, 128–191, and 192–255. The first sub-range is used for the images themselves, without overlays. The second is for those which have the R-bit (6) set. The third is for those which have the A-bit (7) set. Since by definition, the A and R sets are disjoint, no image displayed by REM can have both bits 6 and 7 set, so this range is used for non-image parts of the display, such as the dividing lines between the windows, or the annotations.

Several different look-up tables can be defined using these 4 sub-ranges, and can be combined to give different color effects. Thus, for example, I can display the original images alone, ignoring the overlay bits, such that 0 is mapped to 0 (black) but 63 is mapped to 252 (white). I can show the overlay planes either in solid colors (green for Accept and red for Reject), or transparently combined with the original images. I can use one color to represent error, or different colors to represent different kinds of pixel types—for example, the one most commonly used illustrates correct accept pixels in *green*, misses in *blue*, correct reject pixels in *black*, and false alarms in *red*. I also use spectral color look-up tables to illustrate the levels of the images produced by the coding functions, which are used to get distributions for analyzing the utility of particular operators. The purpose of using all this color is not just to make the system activity more pleasing to watch, but rather to make it more clear to the viewer what the program is doing, what information it is using and how it is making its decisions.

5

Results

This chapter presents the results of the implementation described in Chapter 4, divided into four parts. Section 1 gives a summary of the system's capabilities; Section 2 illustrates examples of the range of different problems solved by the system as part of the program test and evaluation effort. Section 3 describes one complete example run, from start to finish, and Section 4 discusses the issues of efficiency, i. e., how fast is the system compared to a human or compared to an unintelligent machine search?

5.1 Summary of System Capabilities

In order to understand the capabilities of the current system, one first has to understand that a program like REM has different 'levels' of supported operators (see Table 5.1). As we go through the remaining parts of this chapter, it will be important to keep these different levels of operator support in mind. The lowest level concerns support by the languages IC, IA, and MM. This means that it is possible to write algorithms in terms of these operations in the system and have them executed by the image processor. All of the operators and structuring element combinations shown in Table 5.1 are supported to this extent. [1]

Even though one can write and execute a given operation, this does not guarantee that one knows how to determine if it could be a solution to a subproblem. For that, one has to be able to collect data about the effects of the operator on pixels, and to analyze them. In REM, this means that a coding function or routine has to be built to collect data which classifies pixels according to their behavior under the given operation. Distributions taken for the A and R sets on the coded image form the basis of the subsequent analysis. All of the operators in the top two groups of Table 5.1 are supported to this level; the bottom group is not. The top four operators in this latter group (based on thinnings or thickenings with conditional dilation or erosion) are fairly straightfoward; the addition of the grey level operators, though, would require some major extensions to the analysis portion of the system.

Support at the level of the coding functions means that REM has the

[1] See Appendix B for the IA definitions of different classes of these operators.

```
Supported by Search Process:

  Threshold, Identity, Complement
  Infmask-X, Infmask-Y
  Supmask-X, Supmask-Y
  Erode Pdisk
  Dilate Pdisk
  Open Pdisk
  Close Pdisk
  Reduce/Cdilate Square
  Expand/Cerode Square


Supported by Coding Functions:

  Erode Octagon, Square, Diamond, Lines, Vectors
  Dilate Octagon, Square, Diamond, Lines, Vectors
  Open Octagon, Square, Diamond, Lines
  Close Octagon, Square, Diamond, Lines
  Erode/Cdilate Pdisk, Square, Octagon, Diamond, Lines
  Reduce/Cdilate Diamond
  Dilate/Cerode Pdisk, Square, Octagon, Diamond, Lines
  Expand/Cerode Diamond
  Openres/Cdilate Pdisk, Square, Diamond, Octagon, Lines
  Closeres/Cerode Pdisk, Square, Diamond, Octagon, Lines


Supported by Languages:

  Hthin/Cdilate Square, Diamond
  Hthick/Cerode Square, Diamond
  Prune/Cdilate Square, Diamond
  Plug/Cerode Square, Diamond
  Grey-Erode Pdisk, Square, Octagon, Diamond, Lines, Vectors
  Grey-Dilate Pdisk, Square, Octagon, Diamond, Lines, Vectors
  Grey-Open Pdisk, Square, Octagon, Diamond, Lines, Vectors
  Grey-Close Pdisk, Square, Octagon, Diamond, Lines, Vectors
```

TABLE 5.1. List of Operators Supported

capability to solve problems based on these operators, since it has the tools necessary to analyze them. This does *not* yet mean it can do so intelligently—it just means that if it tries one of these operators, and that is the solution, then it will recognize it as such. When I say that REM is *capable* of solving a problem based on a certain operator (which includes all of the examples given in Appendix A), I mean it in this sense—that it will recognize such a solution if it tries to analyze what is in fact the correct operator for the problem.

This level of support is also the basis of the claim that the system has an algorithm repertoire of several thousand possibilities. If we consider that each operator with specific parameters constitutes a separate alternative, then the calculation is as follows: identity and complement each count one. For the threshold band operation, with six bit images and two independent parameters, there are 64x64 or 4096 different alternatives (256^2 or 64K with 8 bit images). Likewise, with images that are 64x64 pixels, the four different position mask operations each give an additional 4096, for a total additional of 16384. (For 512x512 images, this would be increased to 1 million, and so on).

The individual morphology operators listed in the top group in Table 5.1 were restricted to have only eight possible parameter values (for speed considerations), and this gives 64 combinations each or 384. Again, if we increased this for use with larger images, the number of individual possibilities would also increase with the square. Finally, if we take this number and multiply it by the number of morphological operators in the second group (30 without the lines and directional vectors or 78 with them), it is easy to see that several thousand is actually a quite conservative figure. Though the number of different operator and structuring element combinations supported is less than 100, when we consider all of the different parameter combinations for these band operators as distinct alternatives, the numbers quickly climb into the tens and hundreds of thousands. The fact that the system can also find two-step solutions which require complementing the original image before solving the problem with an extensive or anti-extensive operation, would again double some of these numbers.

The fact that one *can* solve problems from a repertoire of thousands is no consolation if we have to try all of them to guarantee a solution. One would hope for an intelligent method which would only look at likely possibilities and ignore the rest, reducing the actual number of possibilities tested to tens or hundreds. There are a number of ways in which the REM system tries to reduce the search which will be discussed in Section 5.4. Some of them are linked to the analysis process (after data collection), but before this one needs others which trim the search space immediately, based on tests which suggest which operators are appropriate for the problem at hand and which are not. The top group of operators in Table 5.1 I will put into this class. They are supported at a higher level (by the search process), because I have found specific tests to apply to the X/A/R images of a subproblem, which distinguish between them under certain conditions and thus allow REM to decide which of them may be applicable. The only such tests currently used for those in the middle group (supported by coding functions and the analysis system), are ones which test for A-Sup and A-Inf conditions, and propose anti-extensive operators in the first case, and extensive ones in the second. While not very specific, this does reduce the search space in half, which is still significant.

At present, the top group includes the trivial operations of identity and complement, the threshold for grey images, the four binary position mask operations, the erosion, dilation, opening and closing by disks, and one extensive and anti-extensive operator on connected components, which selects them based on 8-connected length.[2] I have done some additional work toward the inclusion of the octagon, square, and diamond structuring elements for the basic morphology operations, but this is still in progress. The question of finding good ways to distinguish between the different connected component operators, and especially their structuring elements, is a more difficult problem and will require more work.

In analyzing the capabilities of a program which produces algorithms, there is a kind of hierarchy of levels from the simple to the complex. If we think of a chess-playing program as an analogy, we first would want the program to recognize a mate situation, then to recognize a mate in one move involving one type of piece, then a mate in one move with multiple pieces present, then a mate in two moves, and so on. In a similar way, a program which finds algorithm solutions first has to able to recognize trivial solutions such as identity and complement, which require no search and no real analysis. Then it has to be able to handle problems which can be solved perfectly in one step using one of a set of available IC operators.[3] Most of the implementation effort for the REM system has gone into just this level, of trying to recognize at least perfect solutions over a reasonably wide array of operator types and structuring elements. This at least demonstrates the notion that it is possible to do a formal analysis of a problem and derive the operator which solves it, if everything is 'pixel-perfect'. I also included a number of *different* types of operators (threshold, position, basic morphological, connected component) and structuring elements—to illustrate that the same basic approach can be applied to all sorts of op-

[2]Note that even if we consider just this top group of operators and count up the number of specific algorithms supported by the current implementation, it comes to over 20,000, and over 35,000 when we include the possibility of first complementing the original image.

[3]While they are only one-step IC operations, as mentioned before and as illustrated in Appendix B, they may involve a larger number of algebraic steps, which is a more common measure of image processing complexity.

erators (even non-morphological), which address many different types of image criteria, and everything works in more or less the same fashion. This suggests that with more research we can build algorithm development systems based on formal methods which encompass a much fuller range of the operators typically used.

The next level of difficulty one can imagine in terms of algorithm solutions is the situation where a perfect solution does not exist, but where the number of error points is small enough to be ignorable or within a certain tolerance level specified by the user. There are two subcases here, for morphology operations. In one case, we almost have an A-Inf, A-Sup, R-Inf, or R-Sup condition, but there are a few pixels which prevent this. In these cases, one can try to solve the problem by ignoring these pixels and that is done by modifying the A or R masks to force the desired condition to be true, then solving the revised problem. In these cases, the errant pixels are always part of the set which is *not* going to influence the data collected (i. e., in background for anti-extensive or foreground for extensive), so that ignoring them is an innocuous move. For example, in an 'Almost A-Sup' condition we have some pixels in the image background which create a problem because they are supposedly part of the accept set, A. If we ignore them we get an A-Sup condition, which leads to the examination of anti-extensive operators. These, however, only collect data on foreground pixels; they completely ignore what is going on in the background. This type of situation, and its three other variants, have been implemented in REM.

In the second case, we have an A-Sup or A-Inf condition, but when we collect the data for the A and R distributions, we find that there is a little bit of overlap—they are not completely disjoint. If the error is again within the user-specified tolerance, it is possible to ignore it. In this case the problem is solved by noting the percentages of overlap for the A set and the R set, choosing the one with the smaller percentage, and removing those values causing the overlap from that distribution, effectively ignoring them. Now one has two distributions which are disjoint, and the analyzer can pick parameters to distinguish the two sets, within tolerance, based on them. This case is currently in the process of being implemented.

Once one is capable of solving one-step problems with small errors, the next thing to think about is solving two step problems, first with perfect solutions, and then with errors. Here, too, there are different kinds of cases. One of the simplest kinds of two-step problems involves taking the complement as the first step, and then proceeding as before. The anti-extensive and extensive operators of the REM system are meant to handle A-Sup and A-Inf situations, respectively. They cannot work directly on an R-Inf or R-Sup problem in which the desired pixels are all or mostly in the background. One first has to complement the original image, and this strategy has been implemented in REM. I have also combined this notion with the idea of errors in the masks, described above, to obtain problems which require three search cycles to solve—one to ignore the mask errors, one to complement the image, and the last one to choose the appropriate morphological algorithm. One such problem example is discussed in detail in Section 5.3.

Beyond this, another kind of simple two-step problem to consider is one which the effects of the two operators are completely independent or commutative, as in the case of two masking operations in perpendicular directions. In this case, order of the operators is unimportant, and the problem could be solved as a sequential composition of the two, or by breaking it into two separate parts, and combining results with a union or intersection. The former is simpler and could be implemented fairly easily by having the analyzer or selector recognize that it had two operators offering partial solutions, which were commutative, such that performing both at once could only improve the result of performing just one. This may seem like a very specialized case, but there are such situations, and the ability to catch them and combine the two steps would save a search cycle and thus be quite worthwhile. Note how this type of situation (and others with different particulars) require using some knowledge about the mathematical properties, relationships, and identities between the operators involved, in order to simplify or collapse certain parts of the search. Because of the complexity involved, cases like these will require further study before they can be implemented. As more and more of them become incorporated into the system, however, it will undoubtably require extending the database on the operators to include these kinds of properties and relationships explicitly, rather than burying this in the analyzer or selector code.

The most common types of two-step problems involving morphological operators are the order-dependent ones. Considering only the case where a perfect two-step solution is possible, the key difference here is that now the first step can only obtain part of the solution. One may have a strict containment relationship such as A-Sup or A-Inf, but it may be that while the A and R set distributions are disjoint in terms of the values they take on, there may not be a single *range* of parameter values which can perfectly separate them, because they are somewhat mixed. Alternatively, it may be the case that there is actually significant overlap in their values. Here one has to have the ability to choose a good partial solution, and then set up a subproblem to distinguish the pixels which have overlapping classifications. There are different approaches to such situations, which I mentioned in Chapter 3. One way is to try for a sequential decomposition in which a second operation follows the first to complete the discrimination. Another is to split the problem into two parts, solve each part on the original image, and then combine them.

Two-step problem solving of this type has not been yet implemented in the current system, because it presents several new difficulties. First, one has to be able to make decisions about which partial solutions (which choice of parameters) to select. In general, one would prefer to choose a range which will maintain one of the four strict containment relations—A-Sup, A-Inf, etc., since then only one subproblem has to be generated, of the preferred types. Whether one can discriminate a significant proportion of pixels while maintaining one of these relations will depend on exactly how the A and R distributions overlap. If it is not possible, the best solution may require an attempt to 'split the difference', and then generate two new subproblems to cover the errors in both directions.

While it is fairly straightforward to deal with the analysis part of the two-step problems, this is not the most difficult aspect of them. Since there are many more ways to partially solve a problem than to completely solve it, one also has to have a good way to choose between alternative partial solutions offered by the analyzer. Here too, there are reasonable ways to do this based on the measures of accuracy, consistency, and perhaps cost. The real difficulty presented by two-step problems is in deciding which operators to propose in the first place, so as to avoid analyzing all possible partial solutions. When one is looking at the applicability of an operation as a *complete* solution to a problem, one can be fairly strict about the tests used. When considering the use of an operation as a *partial* solution to a problem, however, what criteria and tests should we use to judge applicability? There are some difficult theoretical problems here, for which few answers exist in the literature; they will require further investigation.

In the section which follows, we will look at some of the problems which the system has solved, from its repertoire of many thousands.

5.2 Examples of Problems Solved by REM

In this section I am going to illustrate some of the types of problems that have been solved by the REM system—a partial list of more than 50 example problems that have been run, and their solutions in terms of the IC and IA languages, is given in Appendix A, to which I will be making frequent reference. Creating an example problem requires the selection of four base images ($X0_i$) and the construction of the appropriate accept and reject masks ($A0_i$ and $R0_i$) for each. The entire process of writing and debugging a program to construct the masks, generating the X0/A0/R0 composite problem images, and executing the problem on the REM system, can take up to an hour or more, so this in itself represents a considerable investment of time.

To make things simple and more clear, I have used the same set of four binary images (or their complements) for all of the binary problems and the same four grey images for the grey level problems. These images are

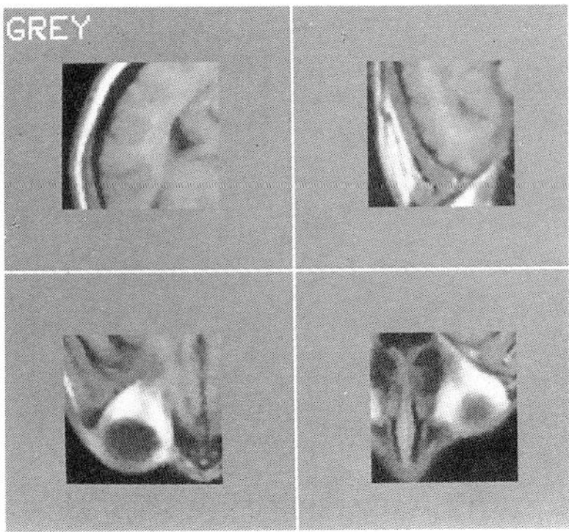

FIGURE 5.1. Original Binary and Grey-Level Image Sets

illustrated in Figure 5.1. Keeping the base images the same in the different problems makes it easier to see the differences in the criteria being used to discriminate pixels in each case, since all that is really changing are the accept and reject masks. The binary base images were designed, in particular, to contain many polygonal objects of different sizes, shapes and arrangements, so that the effects of the different operators being used would be made more obvious. It should be understood, however, that the ability of the REM system to solve the problems that it does depends in no way on the set of base images used (see Chapter 3), but rather on the *relationships* existing between them and their associated accept and reject masks.

In the figures which follow, the four original images for each problem are presented together, with the accept and reject masks overlaid on top. The coloring scheme used is as follows: Pixels which are "on" (white) in the base image and correctly so (part of the accept mask) are shown in green (as hits); those which are incorrectly on (part of the reject mask) are shown in red (as false alarms). Other pixels which are on in the base image but of no concern (neither part of the accept nor the reject mask), are shown in white, since they have no overlay. On the opposite hand, pixels which are "off" (black) in the base image and correctly so (part of the reject mask) are shown in black, while those which should be on (in the accept mask) are shown in blue (as misses). Other "off" pixels which are of no concern are also shown in black. When the base images are grey level, pixels in the accept regions will be shown from blue to green as a function of their brightness (blue if dark, green if bright), while pixels in the reject regions will be shown from black to red (black for dark, red for bright). Their interpretations in terms of hits, misses, and false alarms are analogous to the binary case. Grey level pixels which are to be ignored are shown as their unadulterated grey values.

Careful examination of the colors in one of these image sets tells us something about the nature of the problem to be solved. For example, an image set with just red, green, and black (no blue) pixels satisfies an A-Sup condition (*all* accept pixels are in X), which calls for an anti-extensive operation to be used. A set with only blue, green, and black (no red) pixels satisfies an A-Inf condition (*only* accept pixels are in X), which calls for an extensive operation. Likewise, a set with red, blue, and green (no black) pixels satisfies an R-Sup condition which requires a complement followed by an extensive operation, while a set with just red, blue, and black (no green) pixels satisfies an R-Inf condition which calls for a complement, followed by an anti-extensive operation.

The problems to be illustrated here are divided into seven groups, based on the class of operators involved. The first group (Figure 5.2), includes the trivial operators of identity and complement, as well as the threshold operator which converts grey images to binary. Reading left to right, top to bottom, we have first the identity operation, signaled by the fact that all of the pixels are either green or black (correct accepts and correct rejects). In the next panel is the complement, this time with only red and blue pixels (false alarms and misses). These two problems are solved almost instantaneously by the REM system, as one would rather hope.

The second pair in this figure illustrates two examples of problems solved by one-sided thresholding operations. The one on the left, THRESH_RT, is an 'upper' threshold which selects pixels greater than or equal to some constant, while that on the right, THRESH_LT, is a 'lower' threshold, which selects pixels less than or equal to a constant. Note that the equation for the lower threshold is expressed in terms of the upper threshold and a complement (Appendix A). In the first of them, since the desired pixels are the bright ones, they tend to show up green in the problem display while the darker undesired ones show up as black or dark red. In the second, the situation is reversed, so the desired darker pixels show up more as blue (misses), while the undesired ones are in brighter red tones (false alarms).

The final two panels in the figure show two additional thresholding problem examples, which are a little more complicated. The left one, called IGREY_RT, is similar to the one above it, except that in each image there is a region of 'don't care' pixels around the accept set. These pixels are simply ignored in the tests to construct the algorithm—they show up as grey rather than colored in this display. This problem illustrates the ability of the REM system to solve pixel discrimination problems which do not involve the entire image. This ability is also required whenever it has to solve *any* subproblem other than the top level one. As it turned out, this problem had a solution which was nearly identical to the one above it,

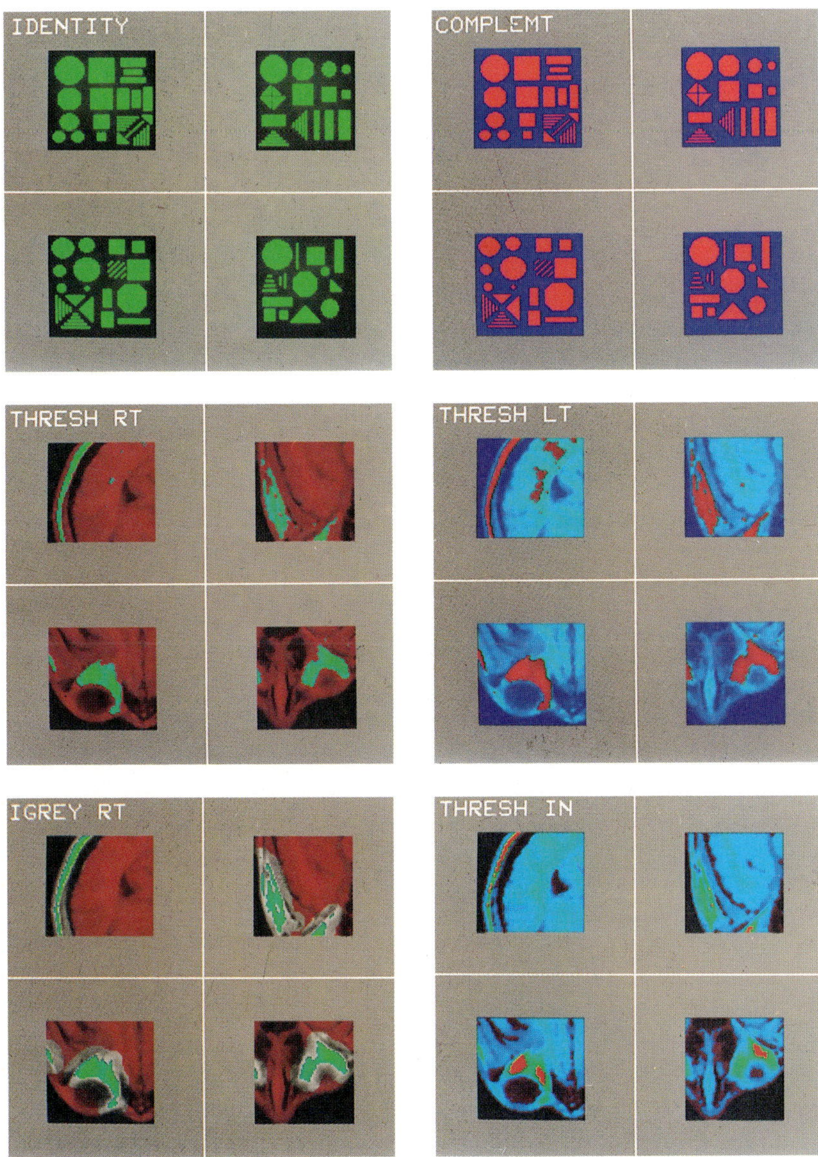

FIGURE 5.2. Identity, Complement, and Threshold Example Problems

since the range of excluded pixels in this problem was not too large. The THRESH_IN example on the right is one which illustrates the concept of a two-sided threshold or band operation—one which accepts pixels which are between two grey values (inclusive) and sets all others to zero. These concepts of one-sided and two-sided ranges or bands will carry over into the morphology-based operations described later on.

Figure 5.3 presents a set of example problems which are solved on the basis of the positional masking operators I defined in Chapter 2. These operators select pixels based on their X or Y dimension addresses. The first two examples make use of the anti-extensive operator (infmask) while the last two use the extensive positional operator (supmask). The INFY_RT example problem is solved by a one sided operator that works in the Y dimension. In effect, it accepted all white pixels in the binary images whose Y address was greater than or equal to 39, and eliminated all others. The INFX_IN example on the right shows a similar operation working in the X direction, but this time using a two-sided inclusive address range to select the white pixels of interest. That these problems represented A-Sup conditions which called for an anti-extensive operation is illustrated by lack of blue pixels in these images.

The second two panels in this figure are variations on the first example. On the left, the IINFY_RT problem is similar to the INFY_RT problem above it, except that once again there is a band of black and white pixels in each image which are to be ignored—they are not part of the discrimination problem. In this case the system produces an algorithm which is somewhat different than that for IINFY_RT. On the right side the CINFY_RT example is one which requires first a complement step before being solved in the same fashion as INFY_RT. This is evidenced by the fact that the images shown all lack green pixels, which indicates an R-Inf condition that should suggest the use of a complement, to change to an A-Sup condition that could be solved by an anti-extensive operator.

The last two panels in Figure 5.3 show problems which were solved by the extensive position mask operators. These operators preserve background (black) pixels based on their addresses, and turn all others to white. The SUPY_OU example made use of an inclusive range of addresses, in the Y dimension, while the CSUPX_IN example illustrates a problem where a complement operator was first applied, to put the problem into a form that could be directly solved by a Supmask-X operator, with an exclusive address range in the perpendicular direction. Note that the solution to this problem is relatively complex—requiring six elementary algebraic steps to accomplish.

Beginning with Figure 5.4, we move into the realm of classic morphology operators. The first ones we will examine are variations on the erosion, an anti-extensive operator used for A-Sup conditions. The first two panels of

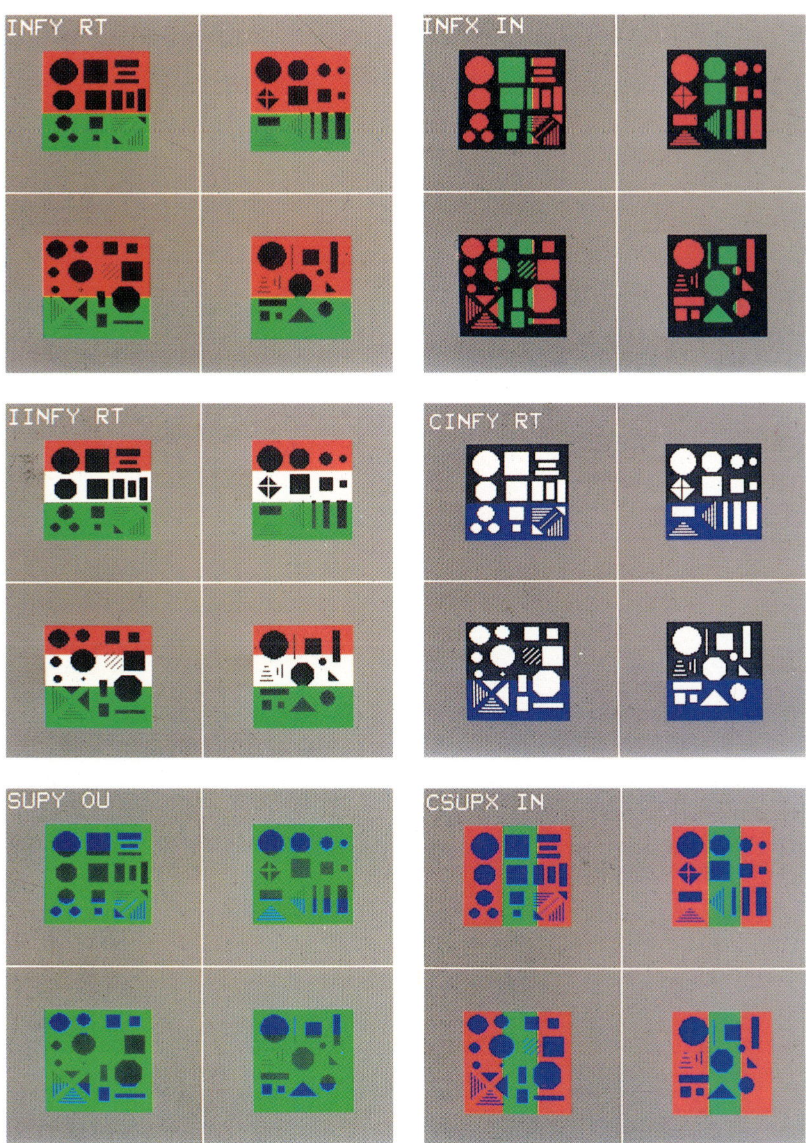

FIGURE 5.3. Position Mask Operator Example Problems

FIGURE 5.4. Erosion Operator Example Problems

this figure illustrate problems that were solved by eroding by disks and by squares, respectively. The first (EDISK_RT) used a one-sided range (all pixels greater than or equal to a certain distance from the background) while the second (ESQR_IN) used a two-sided inclusive range (between two distances from the background). Note the significant differences between the sets to be selected in green, due in part to the different range types, and in part to the different structuring elements being used.

In the next two panels, we see two other problems based on the erosion. In the first, (CEDSK_RT), we have an R-Inf condition which requires a complement, and this then puts the problem into a form identical to that of the EDISK_RT problem above it. On the right side (EDIA_OU) we see the use of a diamond shaped structuring element, and an exclusive range, used to select those pixels which are either close or far from the background. The algorithm for this problem is also non-trivial, considering that this represents just a single IC band operation.

The last two panels of this figure are further elaborations of the first one (EDISK_RT), which both make use of the 'Tolerance' notion. On the left is a problem (TEDISK_RT) identical to EDISK_RT, but which has some additional noise points in the accept mask, scattered in the background of the binary image—these show up as misses (blue). Since all of the other desired pixels can be obtained by operating on the foreground (anti-extensively) and since the percentage of the noise points is less than the tolerance of 2%, the system can solve the problem by modifying it to simply ignore these pixels. Then it becomes nearly identical to the EDISK_RT problem which it already knows how to solve. A further elaboration of this is shown in the last image where we almost have an R-Inf condition except for a few pixels in the accept mask which show up as correct accepts in green (TCEDSK_RT). The system solves this problem by recognizing the near R-Inf situation, modifying the accept mask to ignore these pixels, and then complementing the image, which again gives us the EDISK_RT problem.

The next set of images, in Figure 5.5, are all derived from the dilation, an extensive operator used in A-Inf situations, which is dual to the erosion, and like it, answers questions about distance. The first two illustrate problems that were solved by one-sided dilations, on the left using a disk as the structuring element (DDISK_RT), and on the right using an octagon (DOCT_LT). The former actually illustrates the residue of a dilation, that is, all pixels *less than* a certain distance from the foreground are retained as background, while the latter shows a normal dilation, which accepts all pixels (in the background) which are greater than a given distance from the foreground. In the latter case the foreground is the complement of the usual original binary image. The A-Inf condition is indicated here by the lack of red pixels (false alarms)

The third panel of this group (CDDSK_RT) illustrates an R-Sup condi-

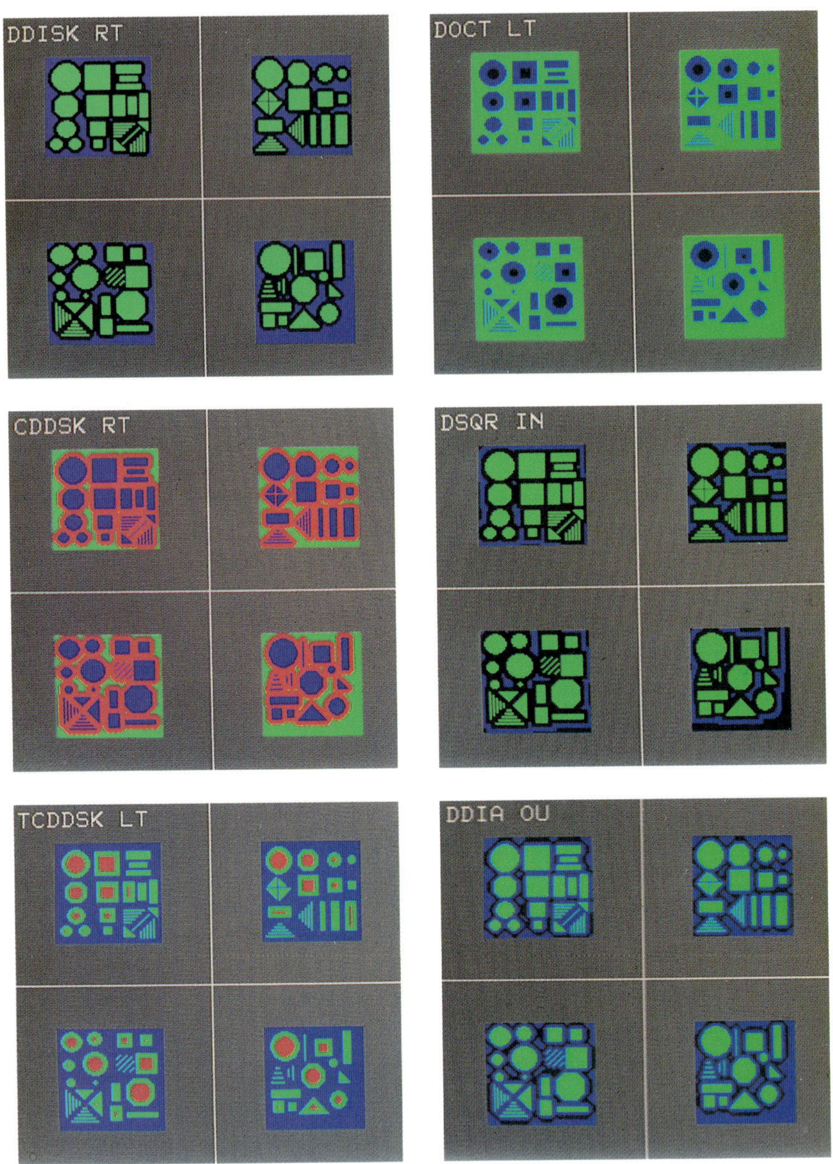

FIGURE 5.5. Dilation Operator Example Problems

tion (no black pixels), which suggests the use of a complement to convert it to an A-Inf, and which in this case produces a problem equivalent to the DDISK_RT example above it. On the right of this, there is an example of a dilation using a two-sided exclusive range, this time with a square as the structuring element (DSQR_IN), and below that, another example which uses an inclusive range, and a diamond shaped structuring element (DDIA_OU). Compare the pixels missing in these two (in blue), with those of the first two panels of this set. Finally, the first panel in the last pair of this set (TCDDSK_LT) is a variation of the one above it (CDDSK_RT), in which a few errors have been introduced into the reject mask (black pixels or correct rejects), but well within the acceptable tolerance of 2%, and such that the desired pixels are based on the basic dilation rather than its residue, as was the case in CDDSK_RT. The system realizes that it nearly has an R-Sup condition here, so it first modifies the reject mask to ignore these pixels, and then complements the image to produce an A-Inf problem that it can solve.

The next group of images (Figure 5.6) are all derived from the morphological opening, which like the erosion, is anti-extensive and used for A-Sup conditions. The examples follow roughly the same pattern as before. The first panel (ODISK_LT) is based on an opening by a disk. Variations on this problem are shown in the two panels below it, CODSK_LT which illustrates an R-Inf condition that requires a complement to bring it into the same form as the first panel, and TCODSK_RT below that in which some errors in the accept mask, shown as green pixels, need to be ignored before the complement step is performed. (This one is also based on a normal opening rather than a residue, as the previous two were). On the right hand side, we have other examples of openings by other structuring elements, using different kinds of ranges. While the first panel illustrated an opening residue by a disk (all pixels whose size class is less than a given-sized disk), the second one (OOCT_RT) shows a normal opening by an octagon (all pixels with size class greater than a given octagonal size). Panel 4 shows an inclusive range, using a diamond-shaped structuring element (ODIA_IN), and below this Panel 6 illustrates an exclusive-ranged opening by a square (OSQR_OU). It is interesting to note the differences between these different sets, and perhaps more so, to see how the effects of the opening operators differ from those of the erosions illustrated earlier.

The following group of images (Figure 5.7) are all based on the closing, an extensive operation used for A-Inf situations like the dilation, but which addresses questions of size in the background areas, much as the opening does in the foreground. The first panel, (CDISK_LT) shows a simple closing by a disk, while the second (CSQR_OU) illustrates the use of a square-based closing, and an inclusive range, to select or maintain background pixels whose size class with respect to a square shape, fell within a certain range. Panel 4 shows an example using an exclusive range with a diamond

FIGURE 5.6. Opening Operator Example Problems

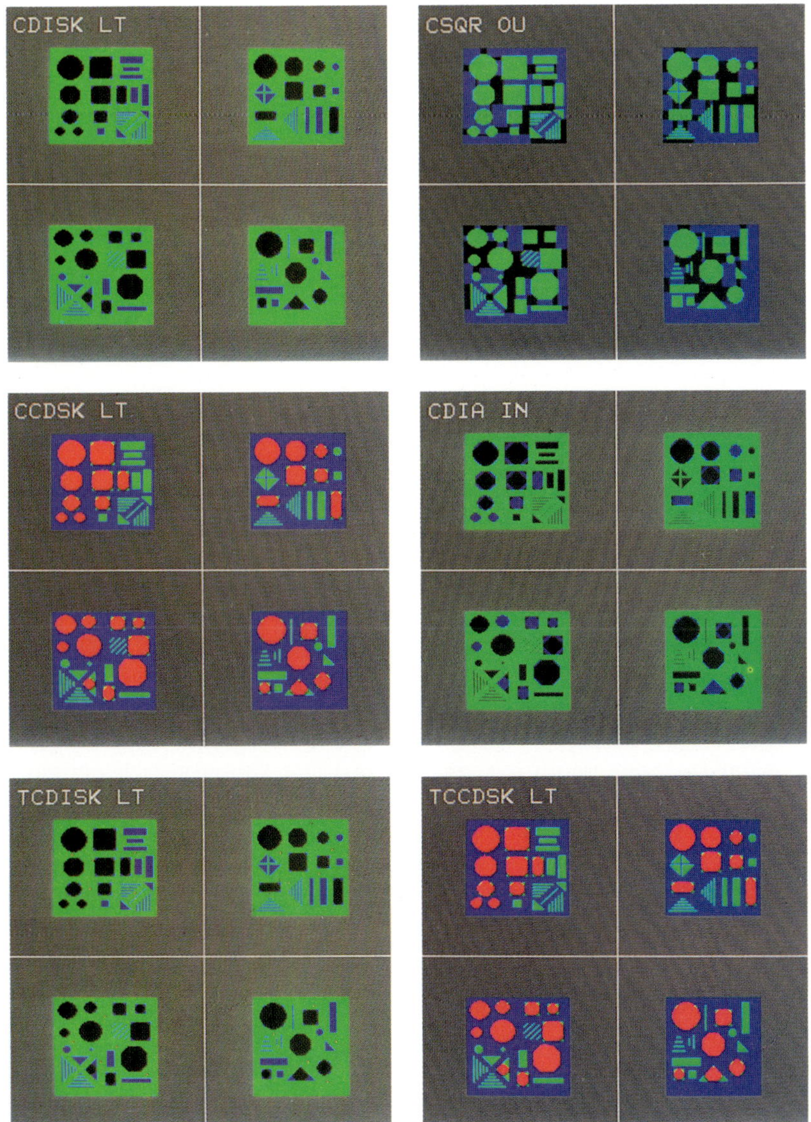

FIGURE 5.7. Closing Operator Example Problems

structuring element (CDIA_IN).

The other three panels in this set are again variations on the first. CCDSK_LT is another case where a complement operation is required to first convert an R-Sup condition to an A-Inf one, which then can be solved by the disk closing. Panel 6, TCCDSK_LT, is the same except that in this case, some noise (black pixels) has been added to the reject mask which has to be ignored before the problem can be solved. Likewise, Panel 5 (TCDISK_LT), is identical to the problem in Panel 1, except that errors of the opposite kind (red pixels) have been introduced into the reject mask which prevent it from being a pure A-Inf condition. Again, the solution is obtained by modifying the masks and proceeding as before. It is interesting to compare the results of the closing operators shown here, with those of the opening and the dilation, illustrated previously.

The final set of images in Figure 5.8 illustrate some operators which are a little different, in that they serve to select entire particles (or holes), which satisfy certain characteristics. Some of them are anti-extensive, while others are extensive. Basically these operators use one type of morphological operator to mark certain particles or holes which are to be selected on the basis of size (width), shape, or length, and then they use conditional dilation or erosion operations to obtain the entire rest of the particle or hole, respectively. The first pair are based on erosion and conditional dilation, which are used to extract whole particles in the foreground, based on their maximum width with respect to an octagon. Panel 1 (EPOCT_LT) is a one-sided residue operation which keeps just those particles whose size is no larger then an octagon of size 3. To the right EPSQR_IN uses a square element and an inclusive range, to select those particles which are either size 3 or size 4, no bigger and no smaller. These two examples illustrate A-Sup situations involving whole particles, as evidenced by the absence of any blue pixels or particles.

In the second pair of panels, two extensive operators are illustrated, which make use of dilation and conditional erosion to select holes in a binary image, also based on size or width. The first, DHOCT_LT, uses an octagonal structuring element and a one-sided range, to select or maintain holes in the original images whose size is at least big as a radius 3 octagon. In the example to the right, DHDIA_OU, an inclusive range is employed to select those holes whose size with respect to a diamond is between 3 and 5. In these cases, we can tell that we have an A-Inf condition because of the lack of red pixels or red connected components.

In the final pair of panels are two problems based on the reduce and expand operators, which are related to thinning (digital 'skeleton') operations on the foreground or background, respectively.[4] These are used with the

[4]Note: The 'structuring elements' shown with these operators are really con-

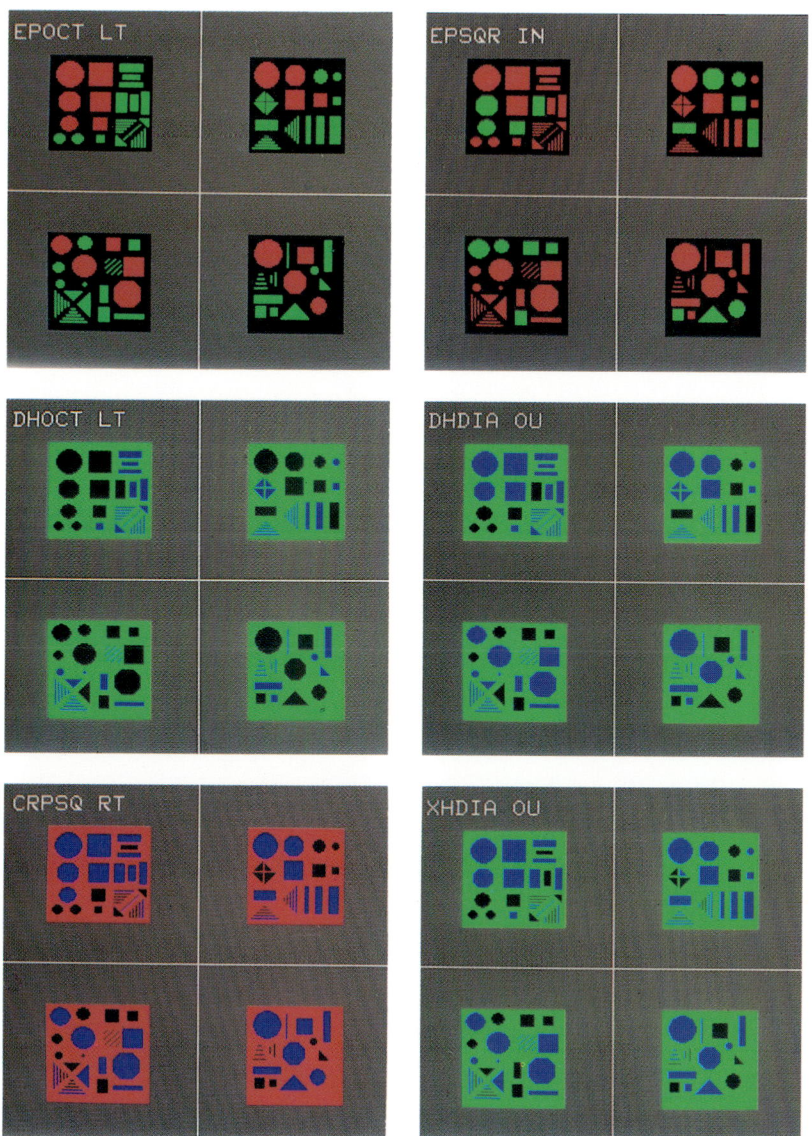

FIGURE 5.8. Connected Component Example Problems

conditional dilation and erosion operators to select entire particles or holes based on their overall length, as measured by the number of steps it takes to reduce their thinned skeletons to nothing. In the first case, CRPSQ_RT, we have an R-Inf condition, which provokes the system to first perform a complement operation, giving back an A-Sup situation. The reduce operator is then used to mark all foreground particles whose length is greater than or equal to 5, and a conditional dilation is used to recover their full extent. This operator was anti-extensive. On the opposite panel, XHDIA_OU, we have an A-Inf condition, requiring an extensive operation. In this case, an inclusive range is used to select or keep those holes whose size is between 3 and 5—all others are filled in. This is done in two stages by using the expand and conditional erosion operators to fill in all holes less than size 3 and then all of those greater than size 6, finishing by taking the union of these two results.

This completes our survey of some of the wide variety of problems that the REM system has successfully solved. As mentioned above, a partial list of the examples that have been solved by the system to date is given in Appendix A. All of these examples have been selected from a space of several thousand different possible algorithms that make up the program's current repertoire. In the next section, I will describe a complete run of the program, from start to finish.

5.3 Example Program Run

In this section, I am going to describe the operation of the REM program in step-by-step fashion as it executes one of the more complicated problems that it is presently capable of solving. A complete transcript of the output produced on the terminal for this run is provided in Appendix C. It would be worthwhile to follow along this transcript as the sequence of events is described in this section. Several illustrations are also presented in this section to show the state of the image displays at important points during the execution.

The problem presented here, TOCDSK_RT, is really a three step problem for the system, like some of those described in the last section. First there is some noise in the masks which has to be ignored, leading to a slight revision of the problem. Then a complement has to be performed, to get the problem into a state more amenable to attack by its operators. Finally the system has to recognize that the resulting problem can be solved through the use of an opening by a disk.

---

nectivity indicators, where 'Square' refers to 8-connectivity and 'Diamond' refers to 4-connectivity.

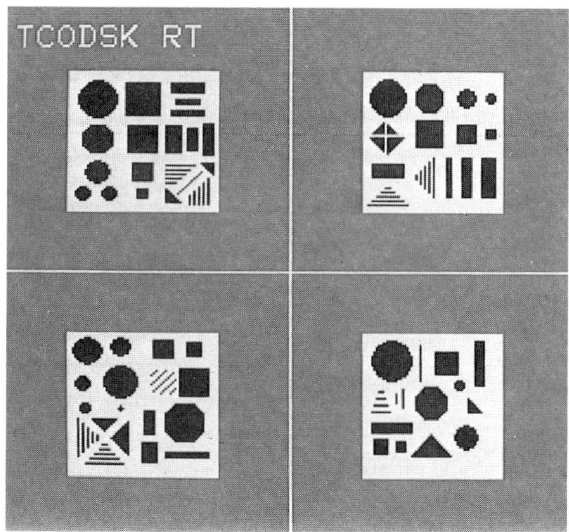

FIGURE 5.9. Original Image Display

The description of this example problem run will follow the sequence of events listed in Appendix C. One note that should be mentioned here concerns the form of the terminal output—in general, mixed case lines represent messages to the user about what REM is doing, or what it has found out. Upper case lines represent image processing code. Most of this is C4PL code (REM prints every line of C4PL code it generates, as it is executed), but in some places there is also MM code (in parentheses with '$' symbols starting the command names), and in other places there is IA code (also in parentheses, but in algebraic form).

Before describing the problem execution I would point out here that by far the largest component of this terminal output is the C4PL command lines—roughly three-quarters of the 35 pages. One measure of the amount of image processing work that REM does in the course of solving a problem is the number of lines of C4PL code it produces. The time it would take a human algorithm developer just to type out the same amount of code listed here suggests that we should be able to make great improvements in speed over the usual methods of interactive image algorithm development, by adopting formal methods.

The REM program begins, once loaded, by reading in a number of image processing code objects and look-up tables. It also declares the display ob-

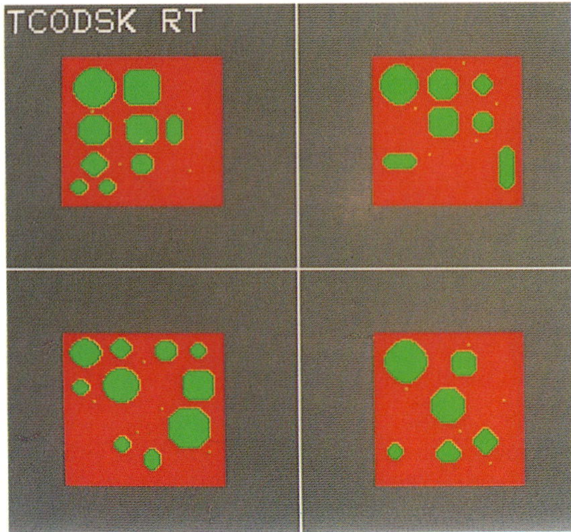

FIGURE 5.10. Accept and Reject Set Masks

jects (Matte and Overview) and defines all of the image objects (X0/A0/R0 and X/A/R image sets) that it will use, as 64x64 windows into these displays. It then generates and displays Matte and Overview. After this initialization, it writes an introductory message to the user, explaining the program briefly and asking for both the name of a problem set, and a tolerance value to use between 0 and 1000, where each unit corresponds to a tenth of a percent.

Once it is given a problem set name, it initializes the main global variables (AST and SPT trees, Agenda, and other lists), and it clears the image displays. Then it begins a sequence in which it illustrates the problem to be solved using the Matte display. It reads in the image files representing the problem to be solved. Each of these contains a six-bit grey or binary image, with two overlay bits, one for the accept mask and one for the reject mask. Figure 5.9 shows the initial display of these images, which illustrates just the four original grey or binary base images, with the overlays invisible. Figure 5.10 shows the next display in the sequence, where the desired accept masks are shown in solid green while the reject masks are shown in solid red. Areas to be ignored, if present, would appear in grey or black and white here. Figure 5.11 involves a different set of look-up tables which are programmed to show hits in green, false alarms in red, misses in blue, and

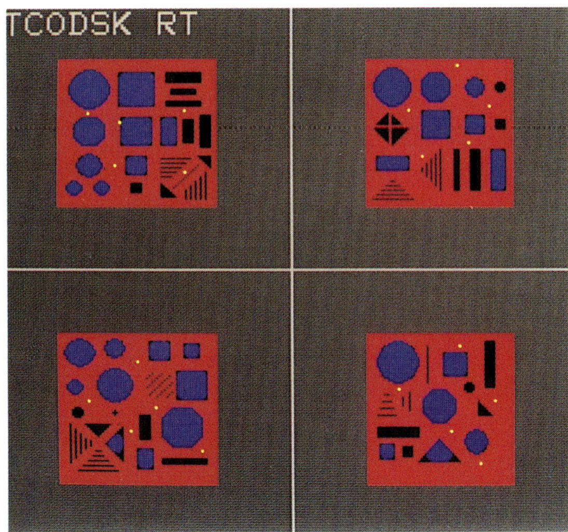

FIGURE 5.11. Accept and Reject Sets Overlaid on Original Images

correct rejections in black. Areas to be ignored are again shown in their original form.

After the display shown in Figure 5.11, the display sequence ends and the problem solving begins. REM first 'loads' the problem into the Overview display, which is used for all of the problem solving work. This is done by taking each image of the Matte display, and splitting it into its X0, A0, and R0 component parts, which are then inserted as individual images into specific locations in the Overview display, as described in Section 4.5. There is about one page of C4PL code in the transcript involved with this loading process.

At the end of loading, the system makes an initial problem evaluation, in which it computes the histograms of the A0 and R0 regions of each of the X0 images. From this it derives the percentage of each A0 and R0 mask in the foreground of each X0 image, which it then draws as coordinates in the evaluation space, illustrated in the large upper window of Overview. Individual coordinates are shown in green; the centroid is shown in red. This display (shown in Figure 5.12), gives a visual illustration of where the problem fits within the evaluation space. From these coordinates the system also calculates accuracy and consistency measures for the problem as a whole, which can be used in developing a strategy to solve it. This

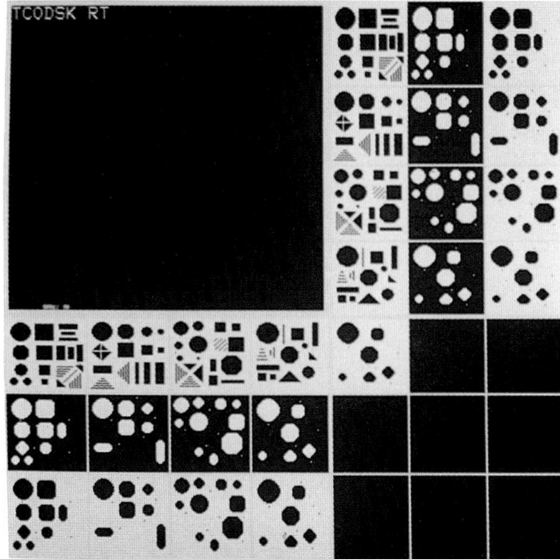

FIGURE 5.12. Subproblem 0 Loaded into Overview

information is placed on the top level SPT node to be used later.

When the initial problem evaluation is complete, the search cycle begins. The agenda is prioritized and REM begins to try to solve the first subproblem (PROB0). It does this by first calculating the images which define the subproblem, and that in turn is done by executing the IA algorithms on the subproblem node which define each X, A and R from the original X0, A0, and R0 images. In this case since PROB0 represents the entire problem, the images which define it are exactly equivalent to X0, A0 and R0 for each instance, so the algorithms are just simple assignments. The transcript shows a page or so of the IA, MM, and C4PL code executed to carry out these steps. These images are inserted into the X, A, and R slots at the upper right of the Overview display, which are used to represent the current subproblem. The state of the display at this point is shown in Figure 5.12.

Now REM is actually ready to try to solve the problem. It chooses two of the images to work with, and invokes the operator expert which begins down its decision tree, asking certain predicates about the X/A/R image triplets and their relationships. The results of these tests are written out to the terminal. In this case REM finds that X is always a binary image, but that none of the preferred strict containment relationships (A-Sup,

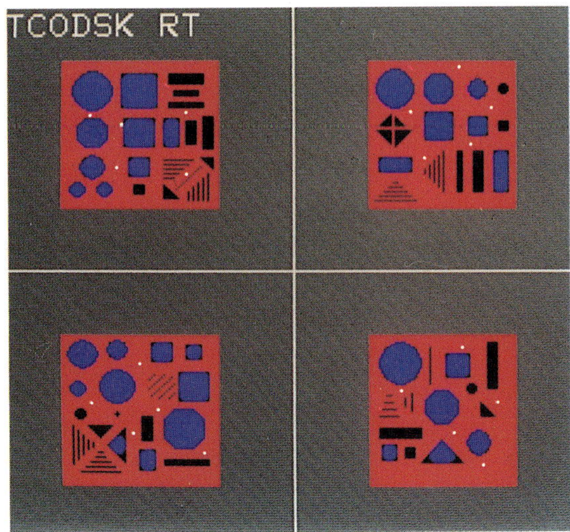

FIGURE 5.13. Subproblem 1 Displayed in Matte

A-Inf, R-Sup, R-Inf) hold. However, in checking the Tolerance parameter against the actual errors found, it sees that it is within tolerance of an R-Inf condition. (All of this analysis is based on the data collected at the initial evaluation).

What this means is that there are a few pixels in the accept mask, which if ignored, would result in an R-Inf condition for the overall problem. By modifying the accept mask in each instance to ignore or remove from it those few pixels which belong to X (the foreground), the problem can be converted to an R-Inf, which is easier to solve.

The operator expert then issues a directive to revise the problem. The data collection step is essentially skipped, and the analysis program follows the directive by creating a new subproblem (PROB1) on the SPT whose accept mask definition will be different. This node will also contain on its blackboard some knowledge of the effect of this problem revision. A new AST node is also created, but it represents the same algorithm as before state except that PROB1 has taken the former place of PROB0. This problem revision is dispatched, updating the trees, PROB1 goes on the Agenda and the system completes one search cycle. When it starts again by prioritizing the Agenda, the only problem left to solve is the new one, PROB1.

FIGURE 5.14. Subproblem 1 Loaded into Overview

It then starts to solve PROB1 exactly as before, by calculating the images which represent it in executing the IA algorithms which compute them. In the page or so of transcript which covers this, note how the algorithms for the A images have been modified to remove any points that belonged to X0. These program modifications were generated by the analyzer and carried out by the dispatcher when it generated the PROB1 SPT node.

When the new subproblem images (see Figure 5.14) have been calculated, REM performs an evaluation of the current problem status by overlaying the new A and R masks on top of the new X images. This is done in Matte, and illustrated in Figure 5.13 with the same color scheme as before.[5] These now represent the new problem to be solved. Note that the previously anomalous points are now in white instead of green, indicating that they are now being ignored.

In Figure 5.14 we see the state of the Overview display after the re-evaluation. The new subproblem images are the same as before except that

[5]Actually, it doesn't have to go through any calculations in this case, because it 'knows' what the new values must be based on the problem modification that was made.

the accept masks are now missing the noise points which they had before. The large evaluation space window on the left illustrates the result of the evaluation of the new subproblem—it now satisfies an R-Inf condition, which is indicated by the red centroid pixel on the bottom line of this window. This condition (which was actually placed on the blackboard of the PROB1 SPT node when it was created by the dispatcher), is noted and reported when the operator expert starts again. On the basis of that condition, it proposes that the complement operator be applied to each base image (X), and it issues a another directive to this end. (The complement operator will have the effect of changing the R-Inf condition to an A-Sup one, which can be attacked more directly by the morphological operators).

Once again the data collection step is skipped, and the analyzer picks up the directive and carries it out by setting up a solution consisting of a complement step and a new sub-problem, PROB2. A new algorithm state node is generated on the AST by the dispatcher, one which replaces the PROB1 link in the algorithm graph with a complement operation followed by PROB2. The SPT is also updated with the new subproblem node, as a child of PROB1. This node will have modified IA algorithms for computing the X base images for the subproblem (compared to PROB1), and it will have some information on its blackboard indicating it is a problem which satisfies an A-Sup condition, based on the known effect of the complement step. PROB2 is finally placed on the Agenda, and this completes the second search cycle. At this point, PROB2 is the only subproblem remaining to be solved.

To solve PROB2, REM once again has to compute the images which define it; this is done by executing the IA algorithms for each one found on the PROB2 SPT node. In this case, however, the algorithms for the A and R masks remain the same—it is the algorithm for X which is modified by complementing. There is another page or so of transcript as the system loads these new images into the current subproblem area.

As before, when these new subproblem images have been calculated, REM does an initial evaluation of the subproblem by laying new A and R masks over the new X base images. This overlaying occupies about one page of the transcript. The result is shown in the Matte display, as illustrated in Figure 5.15. Here we see that the R-Inf problem has been converted to an A-Sup one—there are no blue pixels now; we simply have to remove part of the foreground (red) to get what we want in green. This calls for an anti-extensive operator. In actuality, no computation needs to be performed during this evaluation step; the results are dictated by the previous evaluation result and the known effect of the complement.

In Figure 5.16, we see the state of the Overview display at this time—the X/A/R image triplets of the new subproblem are loaded in the upper right. Note how the X image in each set is a complement of the corresponding X0

FIGURE 5.15. Subproblem 2 Displayed in Matte

image in the lower left group of original problem images. The evaluation space (upper left) also shows a new centroid at the top right instead of the bottom left—this indicates the A-Sup status of the new problem.

The operator expert now starts on PROB2. It picks up the A-Sup condition from the PROB2 node Blackboard, and uses this to begin testing possible anti-extensive operations. It first tests for the applicability of the X/Y position operators which it rejects rather quickly. Then it undertakes a long set of calculations to test four different predicates about the relationship of X, A, and R in the new subproblem (e. g., AllNoMixParts, AllMixBords, etc.). This takes up about three pages. The results of these four predicates allow it to distinguish between particle operators, erosions, and openings. In this case it decides the opening is the correct choice. It then proposes opening by a disk, since that is the only one that the search mechanism supports at this time.[6]

Having selected the disk opening as the operation of choice, the data

[6]In the near future, it will make additional tests at this point to choose between the disk, octagon, square, and diamond-shaped structuring elements, before proposing the opening.

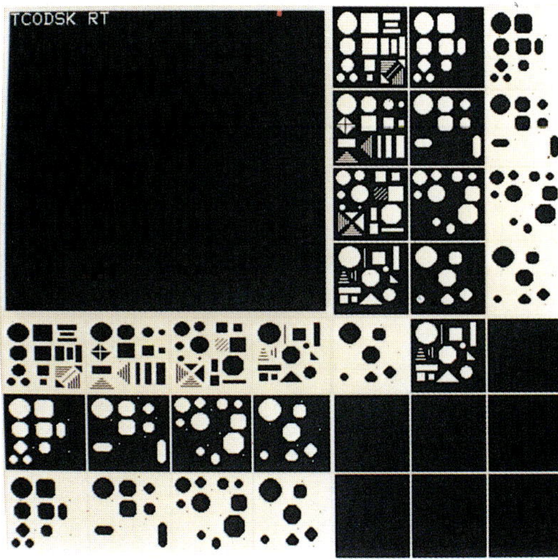

FIGURE 5.16. Subproblem 2 Loaded into Overview

collection phase begins. In two pages of transcript it builds up the coding function for the opening by a disk, for the first image in the example set. This is illustrated in color in Figure 5.17, (large window) where a spectrum look-up table has been used which shows the background areas as black, and the foreground pixels ranging from red (low) through green, yellow, blue and purple (high)—indicating their classification based on that operator.

Next, the grey level coded image is overlaid with the green (A) and red (R) masks, such that the brightness of the grey image underneath is reflected by the brightness of the green or red pixels. This gives a visual sense of whether the two pixel sets (A and R) can be separated by the opening operator. In this case, the green is brighter than the red everywhere, as shown in Figure 5.18. Finally, histograms of the coded image under the green and red areas are taken and displayed as shown in Figure 5.19. This is the actual data from which the system decides that the two sets are discriminable by this operator—if it wasn't a good one, these two distributions would show a large degree of overlap.

In the next seven or so pages, this cycle of coding, overlay, extraction and display of distributions is repeated for the other three images in the sample set. Since this is the correct operator, all of them show disjoint

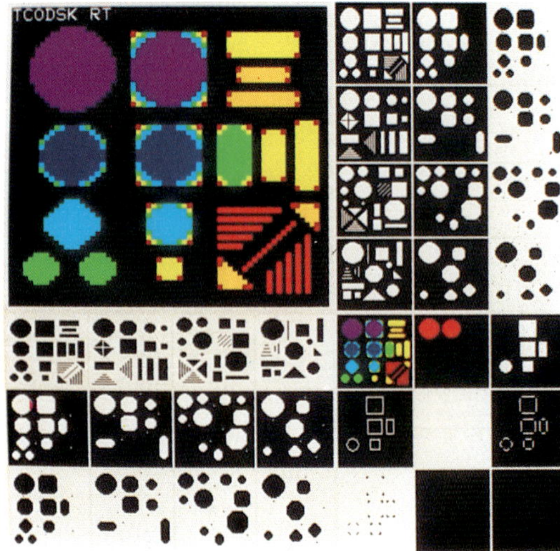

FIGURE 5.17. Grey Level Coding for Disk Opening (Pseudocolor)

distributions for the A and R sets on the coded images. The distributions collected here are stored on the PROB2 SPT node, under the operator for which they were computed.

At this point the analysis phase begins. Here the distributions are checked to see if there is a clear and consistent separation between them. This is seen in the transcript by the display of the 'union of v-grams' for the A and R sets, which is just the union of the binary projections of all the A and R distributions, respectively. The union of v-grams (value-grams) tells us all of the values that the A or R set had, over all of the distributions produced. By comparing these results, we can see if all of the values for the A and R sets were disjoint, or whether they ever overlapped to some extent. Since there was no overlapping in this case, the evaluation of using this operator as a solution to the subproblem, which follows, gives perfect results for the accuracy and consistency measures (1000).

The analyzer packages these measures along with the code needed to carry out this step (i. e., it has to pick the exact parameters to use for the band operator). This is all sent to the algorithm selector, which accepts this solution as it was the only one offered and it worked perfectly. The dispatcher then gets it, and creates a new *terminal* AST node, by replacing

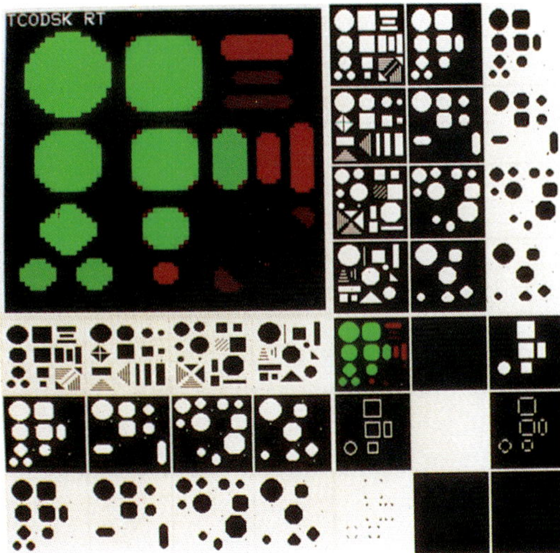

FIGURE 5.18. Overlay of A & R Sets on Grey Level Coding Function

the PROB2 arc of the algorithm with the appropriate code for the band opening. We now have a complete algorithm; no node is added to the SPT, nor to the Agenda. The new algorithm is placed on the NewAlgs list, and is evaluated as solving the problem perfectly (no subproblems are left on it). It is then placed on the Solutions list. We now have a solution on this list and no more problems on the Agenda to solve, both of which satisfy the conditions for termination, so the system terminates the search, and then lists the solution in the IC and IA languages.

Following the announcement of the solution, REM applies the algorithm to each of the original X0 images, and then overlays the original A0 and R0 masks on the results, to 'prove' that the algorithm chosen is a correct solution. As part of the same sequence, a *fifth* instance of the problem image space, which was previously witheld, is loaded into the the $X0_1$, $A0_1$, and $R0_1$ image slots, and the algorithm is applied also to this X0 image and overlaid with the corresponding A0 and R0 masks. The last few pages in the transcript are involved with these algorithm tests. By setting up the system in this way, the program is forced to develop its solution based on an analysis of the first four image sets, and later execute it not only on these, but also on another example of the same problem type which it has never 'seen' before. This constitutes a test of the algorithm on a new instance

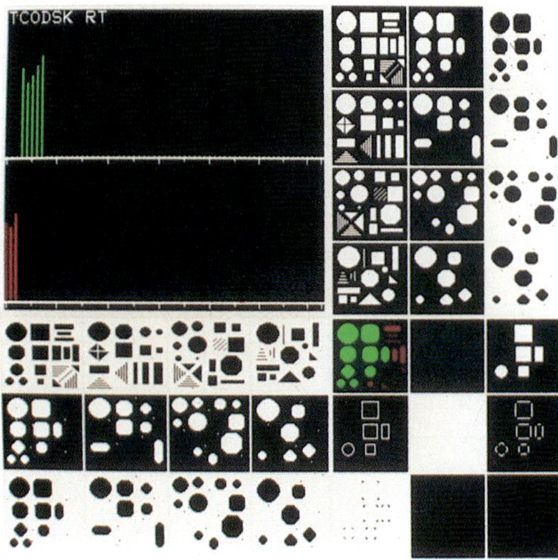

FIGURE 5.19. Distributions of A and R Sets on Coding Function

from the same problem space. If the proposed solution works just as well for the new problem instance it suggests that REM did indeed discover the correct criterion for distinguishing pixels in images of the given problem population.

The final result for this example problem is shown in Figure 5.20—except for the few blue noise pixels which are well within the 2% tolerance, all of the pixels are green or black, indicating correct accepts and rejects, including those comprising the fifth instance at the upper left. This shows that the system has in fact solved the problem satisfactorily. At this point, the top level loop finishes, REM thanks the user for waiting patiently, and then asks for another problem to solve.

That completes this section; in the one which follows we will take a look at efficiency considerations, and try and estimate to what extent REM is performing its algorithm search task faster than a human, and/or faster than an exhaustive, unintelligent search process.

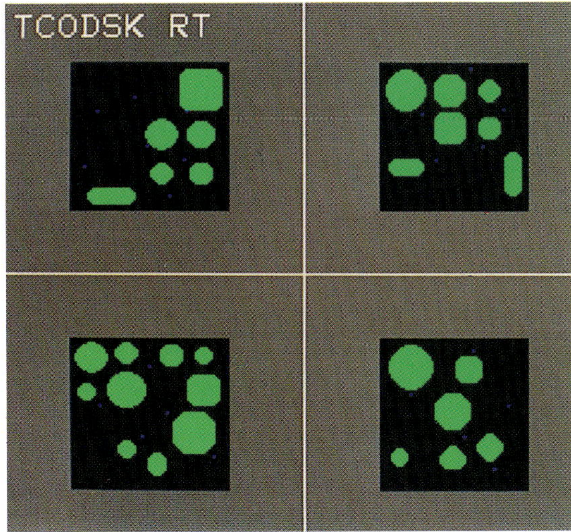

FIGURE 5.20. Final Result of Solution Applied to Original Images

5.4 Efficiency Considerations

Now that we have discussed the capabilities of REM, examined the range of problems that it can solve, and had a look at one complete example run, we can begin to discuss some of the questions about efficiency—how fast does REM solve problems compared to a human, or compared to an exhaustive search process. Unfortunately, the answer to these questions is not a single, cut-and-dried one. There are a number of factors which need to be taken into account before one can come up with a fair picture of what this kind of system offers.

One measure of the amount of work that REM does is the amount of image processing code it produces. In the example described in the last section, this amounted to about 25 pages of C4PL commands, which would probably take a few hours for the average algorithm developer to type out by hand and have executed interactively. Of course this isn't a satisfactory measure, because the human developer can also package up groups of frequently used commands into a procedure which can be called by name and executed as a single command. The human algorithm developer can also avoid some steps, particularly in the data collection, analysis, and parameter selection activities, by eye-balling the images and making rough estimates.[7] In any case, the sheer volume of C4PL code produced to solve the problems represents something roughly equivalent to the amount of code (including packaged code) executed interactively by a typical algorithm developer in a 2–4 hour session, during the initial stages of examining a problem. That is, it represents a fairly sizeable effort, and the fact that it is done here within a fairly short time indicates that formalizing algorithm development does potentially offer some speed advantages.

Along this same line, another measure of the efficiency of the REM system is the elapsed time it takes to solve its problems, compared to that of a human faced with the same ones. REM generally takes between 2 and 8 minutes, depending on the number of steps in the algorithm, and the complexity of the operations examined. Some of this time includes displaying the problem to the user, loading the images at the beginning, and executing the results at the end—activities which are not really part of the search. In addition, this includes time for all of the messages and lines of code that are printed out, all of the program pauses, and all of the image displays and updates. I have also run the system in a production mode where nothing was displayed and no output was produced on the terminal except the final algorithm answer. In this mode most of the current problems could be solved in only one or two minutes. Even though an experienced algorithm developer might be able to see the form of the algorithm solution quickly just by looking at the images and masks, it would probably take him or her at least several minutes to get the right choices of structuring elements and exact parameters which would work over all of the example set images. Just in terms of elapsed time, then, I think the system does comparably well or better than humans, at this elementary stage. Whether this would continue to hold true as more complex problems were considered, would depend on whether the ability of the human to analyze the problem visually more than offset the ability of the machine to make measurements and select parameters directly rather than by the trial and error methods generally used by humans at this time. I won't make any claims at this point; see Section 6.2 for further discussion.

Perhaps the more important efficiency issues concern machine to machine comparisons, that is, how much faster is the REM system in finding a solution than a completely unintelligent search? The answer to this question depends partially on what is meant by an 'unintelligent search'. The REM system does not necessarily cover its entire algorithm space—if it finds a good enough solution during one search cycle it presently stops. However, since each search cycle can expand more than one subproblem node, and since it might be possible for a subproblem to have more than one operator which solves it, it is possible for REM to come up with more than one

[7]REM is a kind of 'blind' algorithm developer, who can only make decisions based on numbers obtained from image measurements.

solution to an entire problem, particularly if they are about the same length. It would of course be possible to change the termination condition, not to stop with just the first solution, but to continue on until some other condition was satisfied—that would require changing the analysis given here.

What this means is that REM is probably most comparable to a breadth-first exhaustive search,[8] which stops at the first level that it finds a solution, rather than a completely exhaustive search which finds all solutions by exploring the entire space, or a completely random one which takes no account of the algorithm length. On the average we can surmise that there is only one algorithm in the space which adequately solves one of the test problems, and that a random or breadth-first exhaustive search is going to have to cover roughly half of the space to find it. This is probably a reasonable assumption for pixel-perfect problems; as we relax the conditions on acceptable solutions, however, the likelihood of having multiple ones in the space increases, particularly for algorithms which only differ by small parameter changes. Given this assumption then, how much work does such an algorithm have to do, and how does this compare to the amount of the space which REM examines carefully, and the amount of additional work it has to do in order to eliminate the other possibilities?

A completely stupid but analogous search would begin by enumerating all of the specific algorithms (including different parameter settings) in the space starting with the shortest. It would execute each one on all of the images in the example set, overlay the corresponding A and R sets on the results, and make an evaluation of the overall accuracy and consistency of the algorithm based on the errors found. It would stop when it found an acceptable result. Since the goal at this point is not to find an algorithm which works best over all test images, but rather one works acceptably on *every* test image, a slightly smarter search process would reject an algorithm the moment it failed on any one image. Since the likelihood of one working by chance is miniscule, in most cases one would only have to look at one or maybe two images in order to reject a bad algorithm.

It is somewhat difficult to do a really fair theoretical analysis of the time or work that such a search would require, and compare it to that required by REM, simply because there are so many unknowns. These include the time it takes to do a particular calculation in Lisp, the time it takes to interpret an IC or C4PL command, the time it takes to set up a Cytocomputer calculation and execute it, the time it takes the Cyto microprocessor to get a histogram, and so on. Also different operators take different amounts of time to execute and analyze, and there are other questions about whether

[8]This too could change with the adoption of more elaborate prioritization algorithms.

to include the time for image displays, and printing information to the terminal, etc.

Instead of trying to come up with a fair equation for all of the 'apples and oranges' involved in the two searches, I decided instead to answer this question empirically by running a few tests, to get at least an idea of how long it would take to perform the exhaustive search. I considered only the operators supported at the highest level (Table 5.1) by the intelligent search used by REM. As mentioned in Section 5.1, this covers more than 35,000 different possible algorithms. I created another top-level program to replace REM, one which removed the entire search loop but retained all of the same lower level machinery for expressing, executing, displaying, evaluating and reporting algorithms.

I then took selected band operators of different complexities (threshold, infmask-X, erosion by a disk, closing by a disk, and one anti-extensive particle operation) and ran them over a subset of their parameter combinations, on just one image, under the assumption that in general only one image test would be required to reject an incorrect algorithm. The purpose of this was to get estimates of how long it took to carry out a test on an average instance of each class. I used the same underlying subprograms as REM, with the same reporting and display regimen, to try to keep these aspects of the two methods as equivalent as possible. Each IC algorithm generated was executed on an original image, $X0_1$, the two masks ($A0_1$ and $R0_1$) were overlaid, and evaluation measurements were made to find the percentages of correct hits and correct rejections on the given image. By obtaining results for how long it would take to run a subset of parameter combinations for one member of each generic class, it became possible to estimate how long a partial exhaustive search would take, using all of them.

The results of these tests showed that the time it took to test a single IC algorithm instance was between 8 and 13 seconds, with the average being roughly 9 or 10 seconds. If we assume that the breadth-first exhaustive search has to look on the average at half of the possible alternatives or roughly 18,000, then this works out to about 50 hours or more than two days. If we compare this to the average 5 minutes that it takes the REM system to search over the same algorithm space, we see that the methods used by REM are significantly improving performance over the unintelligent search.

While it is interesting and indeed striking to see the results of a test like that described above, it is more instructive, I think, to look at some of the ways that the REM system uses to reduce its search, and its overall computational burden. There are several aspects to this. Probably the most important time savings comes from the fact that it does an intelligent analysis of the parameters instead of trying pairwise combinations sequentially or at random. Using the monotonicity properties of the operators to classify pixels and get distributions out, REM only has to perform P image calculations instead of PxP (where P is the range of usable parameters), to assess all parameter combinations for a given band operator. These image calculations are, in fact, where most of the real cost of analyzing image operators comes from, but analyzing them in this way is much, much faster than simply trying all pairs and checking the results. Human algorithm developers often work by trial and error over a limited range of parameter values based on what they see, especially when working with small sets of images. When the number of images in the sample set grows, however, it becomes very difficult to work this way and at the same time make an objective evaluation of different parameter values over the entire set.

I think the value of doing an analysis of the operator parameters instead of testing individual pairs, one by one, has been adequately proven by the evaluation of the exhaustive technique described above. One could hypothesize another, slightly different 'exhaustive search' algorithm, which would randomly propose just generic operators and structuring elements, and then *analyze* the parameter settings in the same way that REM does. In this case, the speedup achieved by the REM search mechanism would be less striking than what appears above—the factor would be more like one order of magnitude instead of two, and the real issue being addressed would be how good the operator expert was at eliminating operator and structuring element combinations, and the amount of time needed to make the tests to perform these eliminations. Since the analyses of the individual operators (especially collecting data) is one of the most time consuming parts of the program, there is still much to be gained even at this level by conducting fast, simple tests that allow one to skip some of the data collections and analyses that an exhaustive technique of this type would have to do.

Turning again to other ways in which REM uses to reduce its search, the whole idea of the band operators making up the IC target language is also related to this notion of efficiently analyzing the effects of operators, because it is through these constructs that REM is able to group together what would normally be four distinct but related image operations, into a single one. These four can then be analyzed simultaneously on the basis of one distribution, instead of separately requiring four analyses, and four generations of the same image data. The compactness of expression realized by using IC as the target language for generating the algorithms, is also important in significantly reducing the amount of work that would have to be done by Lisp in the manipulation and construction of algorithms, were it and the IA language below it not present.

REM additionally reduces its search requirements by analyzing relationships between the X, A and R sets which define a subproblem, in order to determine which operators are appropriate and which are not. For example, if A always contains X (A-Inf) it knows to choose extensive operations, and if X always contains A (A-Sup) it chooses anti-extensive ones. It also has more specific tests for specific classes of operators or structuring elements, as described previously. For example, it knows to use particle selection operators if all of the objects in A are complete particles from X, or to use hole selection operators if all the holes of A are complete holes of X. It can also distinguish erosion situations from openings and dilations from closings, based the different effects these have on the borders of particles or holes in the image.

The real key to the power of the proposal phase or operator expert part of the program, is to have, for every new group of operators added to the search process, a simple predicate or measure that would suggest using that group, or clearly reject it. This is really the only way to avoid having the searches become longer and longer as more capability is added to the system. Moreover, REM also performs these relational tests, not on the entire sample of images, but on a random sub-sample of this set (2 images currently), much as a human operator would. It looks for single failures in the sub-sample of images that would categorically reject a class of operators, or consistent success that would suggest that a class should be examined further. While with the current, very small image samples, this kind of approach doesn't represent too much of a savings, it nevertheless does visibly reduce the amount of processing that REM has to do during the proposal phase, and its importance would increase greatly if images sets of 50 or 100 were used.

Beyond these techniques, the philosophy used in the analysis phase of REM is to reduce search by looking for one-step, perfect solutions first, and harder, partial, multi-step ones later. This way if a complete solution can be found with a simple analysis, it avoids spending what would turn out to be quite a lot of time on many alternative partial solutions, each of which is more complicated to analyze. REM also does simple tests on collected distributions to reject certain operators quickly, before determining whether a real solution is possible. For example, in checking for a complete solution to a sub-problem, it will first check if the A and R distributions are disjoint (an easy test), since this is a pre-requisite to the desired condition that they have separate contiguous *ranges*. If this test fails then the operator cannot provide a complete solution.

To futher reduce the overall amount of calculation it has to do, REM additionally saves distribution information on images (which is relatively costly to generate) if this might be needed later on for other analyses. This avoids their re-calculation. Likewise, it stores the answers to predicate tests on the blackboard of a subproblem node, to prevent repetition of those calculations in the event the same predicate has to be tested again for another reason. When generating a new subproblem, too, it will try to place any predicate information on the blackboard of the new subproblem, which can be determined from the information on the parent blackboard,

6

Conclusion

6.1 Primary Accomplishments

There are several elements of this work which I regard as primary accomplishments or contributions, and which deserve special mention here. Some of these are related to the central theme of the work, the automatic generation of image processing algorithms, while others are related more to the domain of operators I chose to work with, that of mathematical morphology. Additionally there are some lesser contributions, related to the implementation effort, which I want to list at the end of this section.

Beginning with the central theme items first, I believe the most important contribution of this effort is the model that it provides of the algorithm development process. This includes the main data structures used in the search (Agenda, Algorithm Search Tree, Sub-Problem Tree, NewAlgs and Solutions lists), the components of the search and problem solving cycles (agenda prioritization, node expansion, new algorithm evaluation, operator proposal, data collection, analysis, evaluation, etc.), and the sequence of events. While the result seems fairly straightforward, looking at it now, it took several iterations to get it to this stage. I had to study my own behavior as an image algorithm developer and try to capture on paper the sequence of steps I was going through in solving such problems. Sometimes as new capabilities were added to the system, they required a re-thinking and a re-structuring of the the overall search and problem solving process.

The design or architecture of the system as a whole was really formulated to answer the central question of this initial effort, which was: How does one go about successfully developing image processing algorithms to solve set recognition problems? What data is required and what steps does one have to go through, in what order? The fact that the top level form of the program has been stable for quite some time now, indicates that these questions have been answered successfully. This model for solving set recognition image algorithms can now be used as a foundation for building further capabilities and power into the system, which will be discussed in the Section 6.3. I expect it to continue to evolve as more elaborate strategies for solving problems quickly and more complicated and realistic problem types are incorporated into the system.

Additionally there are three specific components of the system which I regard as important contributions—the operator expert, the data collection and the program transform leading to the new subproblem. This will speed up the search for the new subproblem, often obviating the need for the initial evaluation, or for calculating the top-level predicates in the decision tree. Finally, some of the predicate calculations themselves have been constructed so that intermediate image results which are common to these predicates can be shared, rather than repeated for each one.

All of these methods are used to reduce the amount of work that REM has to do in arriving at an answer. That is why it does significantly better than the exhaustive search, and generally performs faster than a human operator, for the range of simple problems that it currently solves. (Another reason for the latter is that it automates many of the processing sequences that a human operator would normally type in by hand). Whether all of these techniques will continue to speed up the search enough to make automatic systems feasible when longer, more complex problems are considered, remains to be seen. This will really depend on how well we can come up with tests for selecting appropriate operators and eliminating inappropriate ones, in the case where a single operator can only partially solve the problem. Before giving up on this kind of intelligent approach to automating image algorithm generation in favor of other brute force or semi-random search techniques, though, we should consider for a moment the alternative costs.

Suppose we are looking for a two step solution. If the number of single operator/parameter combinations is on the order of 10^4 or 10^5, then the number of two-step compositions is on the order of 10^9, which at the current rates of checking would represent literally years and years of search. My own feeling is that stepwise analysis of imaging problems is the *only* way to solve more complex problems at all, within a reasonable time, but we need more research to come up with the decision rules which apply to these more difficult cases.

and analysis components, and the evaluation mechanism. Each of these was also designed to answer a specific question. For the operator expert, it was: How are we to decide what problem solving strategy to pursue and which operators to propose as possible solutions, as a function of the state of the current subproblem? For the data collection and analysis system, it was: On the basis of what information can we determine whether a given operator has the potential to solve or partially solve a given subproblem, and how can we do this in an efficient manner, so as to avoid doing a complete analysis of every operator, especially in cases where a good solution exists among those chosen? Finally, for the evaluation system, the question to be answered was: How are we to evaluate how well a particular operator solves a given subproblem, especially when we consider that performance has to be measured over an entire set of example images?

I believe that within the limitations of the scope of the implementation effort, good answers were arrived at for all of these central questions. Each of these sub-systems had to be invented from scratch, based on thinking about what were the correct answers to these questions for the problem domain. In particular the operator expert, which is still under development and which was the most difficult of the three, required careful consideration of the strategies which can be used in solving morphological recognition problems, and of specific simple tests which can be used to suggest that certain operators are appropriate for a given subproblem while certain others should be rejected out of hand. The knowledge required for these kinds of decisions does not exist in the morphology literature, and it is not easy to come up with such tests by plucking them out of the air.

In addition, the development of the data analysis system required research into the analysis of partial solutions, and on the sequence of analysis that should be done to obtain the best solutions first (as they are the most important and often require the least effort to check), and lesser quality ones later. Much of this work has not yet made it into the implementation because of time constraints and because the hierarchy of problem types suggests that these should be developed and tested only when one is sure that more simple types work correctly. I do, however, regard this as one of the major contributions of this effort, because it too had to be invented, and took considerable time to develop since there are a number of subtle issues involved, some of which are not yet completely resolved.

On the morphological side, there are also two or three things developed under this project which deserve mention as important contributions. The first and most important of these was the concept of the band operators based on underlying morphology operations. I did not completely invent this notion—Serra and others who work in the field are quite familiar with the concept. What I think I did do was to recognize it as a very central key for the goal of developing morphology algorithms automatically. This comes about for two reasons. First, the band operator concept allows one to group a set of four related operations, all based on a single basic underlying morphological operator, into a single unit. They can be analyzed together based on a single distribution, which is much more efficient than trying to test each one separately. Second, the band operators speak to more conceptual notions about images—unlike the individual steps which make them up, the band operators tell us something meaningful about the pixels of an image—they have a more direct relation to the discrimination criteria which we use to express differences between groups of pixels in an image.

In addition, the band operators, as defined in the IC language, represent a remarkably compact way to express some very complex sequences of operations—in some cases a single band operation represents as many as a dozen elementary steps in the same algorithm expressed algebraically. This represents a tremendous increase in power, because it would take much longer to develop the same algorithm in terms of the individual steps if we were forced to do it that way. Thus, while the present REM system only solves problems that are a few steps long in the IC language, they are somewhat longer when measured in terms of the elementary steps required to execute them, which is the normal measure of complexity used in the field. If this work has demonstrated one thing, it is that the band operators as defined here represent the appropriate level for thinking about solutions to image algorithm development problems, and likewise for generating such solutions automatically.

Having recognized the central importance of the band operators to the problem of developing image algorithms automatically, I pushed on this area in ways that led to additional developments which I regard as important. First, I developed band operator definitions for a large number of purposes—these were discussed in Section 4.1, and are listed in Appendix B. Many of these definitions, especially for particle and hole operations based on skeletons or residues, were fairly complicated and took quite a while to develop. Some of them have not been previously defined, making this the first time they have appeared in the image processing literature.

Second, I noticed almost from the beginning that there was a problem with the traditional definitions of residues for extensive operations. This was described in Chapter 2, but basically the problem was that while residues for anti-extensive operators were anti-extensive, the residues as defined for extensive operators were not extensive, meaning that we couldn't handle them in a symmetric way. Moreover, while it was easy to describe the meanings of anti-extensive residues in terms of the effects of the underlying base operator, this was not so true in the case of the extensive residues, as defined. Finally, it became apparent to me that the concept of using band operations as the foundation of the REM program's analysis of operators was not going to work for extensive residues if something didn't give. I eventually decided that the definitions of these operators had to change to make them consistent and dual with the anti-extensive cases, and to give them meanings which were in line with the meanings of the base operators from which they were derived.[1]

I managed to make these changes first for the binary residue cases and then for grey level residues, which require slightly different formulas. Once this was done, it was a much simpler matter to define the inclusive and exclusive range versions, and then to use these same principles to define extensive band operators for the more complicated cases described above. Having done this, everything else on the analysis side fell into place rather quickly, and the idea of using band operations as the foundation for all of the data analysis done by the system was firmly intact. I am now convinced that these new definitions are the correct ones, and the ones that should be used generally in the field. I thus regard this as a major contribution to mathematical morphology which has resulted from this work.

A related contribution, also discussed in Chapter 2, and for which I have written a paper [Vogt 1988a], concerns the problems created when trying to use digital disks when constructing operator distributions for openings and closings. Unlike digital octagons, squares, diamonds, and lines, digital disks do not have the property necessary to guarantee an unambiguous classification of every pixel in an image when opening or closing—the successive results generated by using larger disks do not always satisfy a strict inclusion relation. This result is not commonly known.[2] I was not aware of it and did not expect it when I started to develop the band operator framework for analyzing operations in the REM system. I did not remember hearing of this problem in my discussions with Serra's group while in France, perhaps because in theory they use Euclidean disks and in practice they use hexagons or dodecagons, all of which do not generate this problem. It is an important problem because digital disks are the ideal structuring elements to use for many types of situations due to their isotropy. I was surprised when the REM program failed to solve the disk opening and closing problems correctly, missing only by a few pixels. After a week or so of investigation, I realized what was going on, and after another week, came up with an acceptable resolution to it, which is reported in the paper. While in some sense this is just a technical detail, I feel it is important because of the general utility of digital disks in image processing work, and because it required developing a slightly different method for collecting data in these cases, compared to the majority of other cases—one which deserves to be known by practitioners in the field. The solution I proposed involves resolving ambiguously categorized pixels by preventing their re-classification

---

[1] It is this sense in which I say that while I did not invent the concept of morphological band operators, I did make a contribution to it, because without redefining the extensive cases, the concept doesn't work as well as it should.

[2] There are a few references to it in [Serra 1982]; in particular, page 192 and Chapter X, Section C.1.

once they have been initially been classified. This in effect gives priority to the smallest class when there is more than one from which to choose. In most real images the percentage of these ambiguous pixels is rather small, and they tend to be scattered about, rather than contiguous or tightly clustered.

A final important contribution in the area of morphology coming out of this work concerns one of the other main subjects of Chapter 2, and that is the criteria addressed by different morphological operators. In studying these operators, I tried to come up with English explanations of their meanings in terms of the visual aspects of an image. This kind of explanation exists only to a very limited extent in the morphological literature, and when present, is usually very general. I developed very detailed explanations of the effects of specific operators, not only to improve my own understanding of them, but also to understand the circumstances for which they should be used. Some of these results were reported in Chapter 2, but many of them were not. In the initial stages of the thesis effort, I researched a number of additional operator classes in detail, which I eventually decided not to include in the implementation, and which are not discussed in Chapter 2. I hope to be able to publish these results at a later time, because I think it would help in training students in the field to have detailed explanations of the precise effects of some of them.

That completes the description of what I consider to be the most important contributions of this effort. Before leaving this section, however, I want to mention several other components of the work, which, while not major contributions, were either important to the implementation effort or required a substantial amount of time to generate. The first of these was the Lisp/Cytocomputer interface. This was an extremely important development for the overall project. First, it satisfied the overall goal concept of automatically generating image processing algorithms in Lisp and running them immediately, interactively on the image processing machine, and also in using the machine in the analysis process for developing the algorithms to program it. Second, it made it possible to get turn-around times in the course of solving problems that were reasonable. The system could solve its problems in ten or fifteen minutes of elapsed time instead of the hour or more it might have taken on the Vax alone. This made it possible to test the program faster, and to run many more types of problems than I would have been able to do otherwise. With the recent addition of the new display hardware, which reduces by a factor of 10 the time it takes to display an image, the system solves its problems in about five minutes, and I am able to illustrate many more of the steps the system goes through which I had suppressed before due to time considerations.

A second important component was the development of the IA language for expressing morphological algorithms algebrically in Lisp. This development made it much easier to express algorithms to be executed, than it would have been if I had worked directly with the C4PL language. The uniform syntax of IA compared to C4PL, the fact that multiple steps could be expressed as a single statement, and that arguments were variables which could be instantiated with real images at the time of execution—all made it easier to do the manipulation and constructions needed to make the algorithm generation process work smoothly. Moreover, this language embodied the graph approach to the description of image algorithms, which I feel is crucial to the development of systems that combine image processing and artificial intelligence. Though from a computer science standpoint it was perhaps not a great accomplishment, it was to me personally because my training has not been the area of language development, and thus the problems of recognizing certain constructs, allocating image registers dynamically, and generating executable sequential code were rather new and challenging to me.

A third important component of the implementation effort concerned the development of the coding function algorithms, which were used to generate data for each individual class of operator supported by the REM system. Like the IC definitions of some of the band operators described above, most of these algorithms had to be invented, and some of them, like the ones which select particles or holes based on their skeletons, or on the roughness of their surfaces, were not simple to generate.

Another element of the REM system which deserves mention and which also took considerable time to create, was the display implementation. The display implementation includes the Matte and Overview displays, and the look-up tables used to illustrate different things about the images and the problems. While it was of course possible to run the entire REM system without displaying anything (which makes it go considerably faster), for debugging purposes and in order to make it more interesting and educational, I had to decide how I could display all of the images that the system used and generated, and how I could use color to illustrate important facets about the problems, such as what were the accept and reject sets, what was the current subproblem definition, what did the coding functions do to a given image, and what were the distributions by which it decided to use or reject a given operator. The decisions to use 64x64 images with 6 bits of grey and two bits of overlay, as well as the decisions about how to arrange the images, and what color schemes to use to illustrate certain things, all took considerable thought and time which were not directly related to the central goal of the work. In the end I think it was time well spent, because the resulting displays do a very good job of illustrating all of the important image processing and analysis activities being performed by the system, as some of the photos in this book illustrate.

Finally, another fairly time consuming element of the implementation effort was the development of test image sets and their execution, which are described in some detail in Section 5.2 and listed in Appendix A. In all, roughly 60 different problem sets have been generated and executed, to illustrate the range of different situations which the system is able to solve. This in itself represented well over 100 hours of effort.

That concludes what I regard as the most important accomplishments of this initial effort toward generating morphological algorithms automatically. In the next section, I want to examine some of the more philosophical issues as well as potential criticisms suggested by this work.

6.2 Summary and Discussion

In these pages, I have described a formal approach to attacking the problems of image processing algorithm development. I have described a theory or model for this kind of activity, and have implemented a system which embodies that theory to a limited extent. I have shown that the approach described here is truly a formal one, in the sense that the search process used is completely independent of the precise input data given to it, and it does not miss any solutions within the scope of its repertoire. Put in other terms, it does not make guesses—it uses mathematical knowledge about the domain of interest to do an intelligent job of finding algorithms. In addition, I have shown that this search mechanism is efficient, in the sense that it finds its solutions generally faster than humans, and much faster than would an equivalent exhaustive search technique, over the range of algorithms within its repertoire. Finally, I have demonstrated the system on a fairly wide assortment of problem types.

While the implementation effort described here focussed primarily on a subset of classical morphological operations, the same basic methods would be applicable to many other operators, including non-morphological ones. I believe very strongly that the increased use of more formal methods into the image algorithm development process has much to offer algorithm developers. As I have already stated in several places, the study and formalization of the algorithm development process has three major benefits—it first helps us to understand better the operators we are using and when we should use them; it promotes the development of more powerful analysis tools which can improve the efficiency and motivation of algorithm developers, as well as the quality and robustness of algorithms; and finally it promotes the development of systems to help us actively manage the huge task of searching through the algorithm space, for that arrangement of operators which best solves a given problem over many example images. Since we know already that many of the tasks involved in algorithm development are ones that machines can perform quite well, there is really no good argument against proceeding more and more along this direction. The major problem is to find a way to put our image processing expertise, and the mathematics behind this, into computer-usable form.

One might reasonably ask at this point, to what extent do I see such programs taking over the role of algorithm developers in the future—how good do I think we can make such programs? Related to this is the question of how I think the theory and system developed here could 'scale up' as more and more operators are introduced, and as the system is faced with longer and more complex problems.

My own view is that we will probably never have algorithm development programs which rival expert human algorithm developers—at least not for a while. The human eye and brain have abilities to focus on the salient aspects of problems, to bring only the relevant knowledge to bear on them rather quickly, to use intuitive judgement based on years of experience to decide between alternatives, and to integrate the many elements of a problem in ways that we simply cannot yet reproduce very well in a machine. I think one element of this is that the human visual system appears to do quite a bit of built-in manipulation of the sensory data it receives, over many different channels. Primitive information about size, edges, direction and orientation, closure, brightness, color, and so on, are there from the beginning, pre-computed [Marr 1982], whereas a system like REM has to compute these things sequentially, in order to have data with which to work. For the human the problem becomes more what to focus on and how to combine these elements, while for the machine it concerns which data analyses to carry out in the first place.[3] On the other hand, the machine can perform many activities much faster and more reliably than humans, so there exists the potential for improving performance beyond that of humans if we can choose the right subtasks, and put our mathematical and experiential knowledge about them into computer form.

Whether or not we can ultimately build programs which can develop image algorithms as well or better than human experts, for the present we are faced with the fact that much of a human algorithm developer's time is wasted in doing things that a machine could do much better or faster. There are great gains we can make in improving the efficiency of algorithm developers and in relieving them of a certain amount of drudgery. I think that we *will* be able to have machines solve relatively simple real-world imaging problems in the not-so-distant future. It may also be possible to use such programs as assistants to humans—for example, filling in details of an approach sketched out by a human algorithm developer. Indeed, the current REM system could be adapted to such a use by simply making it

---

[3]It might be wise to consider the possibility of an image processing system which automatically computed many of the most important visual criteria measurements simultaneously in parallel, and then selected among these outputs the ones which were most discriminating as a post-processing, attention-focussing step.

possible to initiate the global data structures, particularly the Algorithm Search Tree and Sub-Problem Tree, with a partially developed algorithm state and set of subproblems to solve.

For me the real value of working on this kind of rather challenging problem is that it helps us learn how to work better so that we can ultimately do more. By studying the operators we use, we come to a better understanding of when and how to use them. By studying how we develop algorithms, we can make this activity more efficient, and actually automate parts of it. By studying how we manage the search and evaluation of algorithms, we can come up with ways for the machine to help us, and thereby increase our own effectiveness and power against problems which are combinatorially formidable. The expectation is not to be able to solve all such problems automatically, but rather to be able to do a lot better than we can do now, through the use of tools that help us get to solutions more quickly, or help us get to better solutions, by helping us to explore the search tree faster and more efficiently.

Indeed, whether or not we ever arrive at programs which can solve complicated real world imaging problems, just attempting to do so offers many possibilities for improving our own capabilities in this domain. Since in general the problems are very difficult, I believe that anything we can do to aid the current situation will be welcomed by those in the field. There is so much already known, and so much that is left to be known about the operators, about how to use them with each other to form meaningful units, and about how algorithm development is really done by humans, that the task of integrating all of this into a coherent framework is one that is likely to take many years. The effort described here represents only a beginning towards that end.

I think that one of the great benefits of constructing a system like that described here as a computer program, is that it allows us to model the algorithm development process as best we can, incorporating all the expertise that we have or can find from whatever sources we have available. We can then test it, see how it performs, and offer corrections or additions to it based on the behavior it exhibits. Over the course of time this allows us to turn raw, scattered, unshared, often forgotten expertise into productive, concrete, re-usable expertise that stays around and is transportable.

Concerning the issue of scaling effects, there is a question as to whether or not the type of system implemented here will continue to work well as many more operators are added, and as the system has to address problems which are larger and more complicated. My own feeling is that scaling is quite possible, but that it will not be a simple extrapolation, or a simple matter of augmenting the operator database, languages, coding functions and operator analysts. First of all, each new class of operators brings with it a new set of idiosyncrasies which have to be integrated into the rest of the system, and each one is used somewhat differently which requires new knowledge to be developed and incorporated. More important, however, is the fact that as we add new operator classes, and capabilities for handling problems which require more steps, our expert system has to become smarter and smarter at differentiating the alternatives, coming up with tests which eliminate fruitless paths quickly and focus just on possible solutions. The knowledge for making these kinds of decisions quickly and intelligently will have to be developed—for the most part it is not documented anywhere.

In addition, while the fact of having built a model of the process and having implemented several classes of operators, may make it easier to add new operators rapidly in the future, the other side of it is that the move from complete solutions to partial solutions, and from algorithm graphs that have a few arcs to those which have many, is a big jump, because all of a sudden there are many more ways to solve a problem which need to be either rejected or given further consideration. This is not to say that we cannot handle these situations automatically, but they do present new difficulties. There is much to be learned and much painstaking effort that will be required, for us to accomplish such goals. Every improvement in operator capability and in the ability to handle more complex problems will come about only through a full understanding of those operators and the new classes of problem complexities being added. In addition, we will also need to have a better understanding of how we go about solving the more complex problems in this domain ourselves. There are no simple shortcuts, but there is great potential for making significant improvements over our current levels of efficiency and capabilities.

Turning next to the issue of criticism, aside from potential complaints about this work that not enough operator classes were considered, or that the expert system was not far enough developed, or that the difficulty of the problems solved was not representative of real world situations (see Section 6.3 for further discussion)—there is one possible criticism of it that I want to address, because it is more substantive than the others, (which I regard really as questions about how much could be done in a given time within given constraints). This criticism concerns the empirical nature of the method.

The system described here attempts to find algorithms for image processing problems based solely on empirical techniques. It makes measurements on images, collects data, analyzes them, and tries to find a solution which it knows will work based on its mathematical knowledge of the domain. The human algorithm developer, however, usually has more than this at his disposal, in the form of *semantic knowledge* of the problem domain. This may include a knowledge of the sensor characteristics, knowledge of the radiation being used to generate the image, knowledge of the physics of the object being imaged, or knowledge of constraints about how that object will appear and in what contexts. All such information can be extremely valuable in whittling down the search space to manageable proportions. In fact, if this information is known in sufficient detail, it should theoretically be possible to generate an algorithm directly, without any search, simply from a knowledge of the criteria which consistently discriminate the object of interest from the background, combined with expert image processing knowledge of the operators which address those criteria.

In reality, we usually do not know the physics, sensor characteristics, object and background characteristics sufficiently well to be able to generate algorithms in this fashion. Particularly when we are dealing with natural rather than man-made objects, our knowledge of these elements is often not enough for us even to generate accurate models of the objects under study, much less to determine algorithms to recognize them without an empirical compenent.[4] For most imaging problems of any real difficulty, empirical testing is an essential part of the algorithm development process, and that is what the REM system described here was designed to address. The point here, however, is that a second essential component—semantic knowledge of the problem domain—may be just as important or more so if it exists, and this is something that the REM system does not address. Even when our knowledge of the problem domain is not sufficient to determine an algorithm *a priori*, we may nevertheless know enough to eliminate a large number of possibilities based on knowledge that they could *never* occur, or to focus on the most promising ones based on knowledge of what is *likely* to occur.

My response to this potential criticism, which I think is a quite valid one, is that the goal of this research effort was to develop a better understanding of the empirical component of the algorithm development process, which is in any case a necessary component of the complete process, and for which much work remains to be done before we have a good handle on it. The theoretical component is in a sense perhaps simpler, in that once the distinguishing criteria are known, it is relatively simple to construct an algorithm from operators which address those criteria, based on exactly the same expert knowledge which is required by an empirical system like REM. The real problem posed by the theoretical component is more one of deducing discriminating image criteria from a knowledge of physical events (a very difficult problem), or barring this, at least having a way to represent knowledge about the problem domain and the constraints on

---

[4]Even with one, a further common problem is that we never seem to have enough data, in the form of image instances, to be sure that the assumptions we make on the basis of observation are sufficiently stable to serve as reliable criteria for an algorithm, especially when we want to use that algorithm in new, potentially different situations. We often cannot make up for our lack of knowledge about the problem situation by collecting more data, because it may be very expensive to collect.

it such that this information can be used easily to reduce the algorithm search space. This is more a matter of semantic knowledge representation than of algorithm development, *per se*, and while I regard it as a necessary component of any system which is to rival human performance in image algorithm development, I felt that it was reasonable for the purposes of this effort to ignore it.

So much for my thoughts on where this type of work may lead in the long term future. In the final section which follows, I am going to look at some near term future directions for this work and for the REM system—improvements I would like to make, extensions to its capabilities, and additional research that I would like to pursue.

6.3 Future Directions

At the end of any extended research project, there are always some things which could have been done a little better, and some which have to be left to future research, or other researchers. While the REM program, as implemented, represents a good first step towards the goal of having systems which can do expert level algorithm development for morphological set recognition problems, there is much more work to be done and many more improvements to be made. In this section I want to list some of these areas briefly.

One near-term goal is to bring the rest of the operators which are currently supported by the languages, up to the level where they part of the intelligent search process. For some this means adding coding functions in order to collect data for them, as well as inventing tests in the initial proposal phase (the operator expert) which will suggest or reject them as potential solutions. In particular I would like to add linear structuring elements (lines and directional vectors) to the erosion, dilation, opening, and closing operators, because these elements will allow the system to address some new and interesting kinds of problems.

Related to this goal is the desire to extend the system to handle grey level morphological operators—first ones which take a grey level image and put out a binary image directly, and later ones which simply do grey-to-grey transforms. Grey level operators are more complicated than binary ones, more data has to be collected to analyze them, and the analysis itself is somewhat different. The introduction of them will require significant extensions to these parts of the program, as well as to the operator expert.

Later on I would like to add still more operators to the system, including morphological ones that deal with other types of criteria such as connectivity, adjacency, pattern, and texture, as well as smoothing and filtering operations coming from both the morphological and linear domains. The smoothing and filtering operations, especially, would require the introduction of some new issues and questions into the system, for example, on what basis do we decide that a particular filter is doing an acceptable job? The answer to this question is not so cut and dried as that of trying to distinguish two sets of pixels.

Another major goal is to improve the capabilities of the system to handle more complex problems, eventually moving to more realistic ones where noise is a bigger issue, and many more steps are required. While the present system has some capability for handling noise and multi-step problems in simple situations, much more remains to be done in this area. The first goal would be to have it be able to solve problems where two different band operators must be used together to solve the overall problem; that is, where the first step used could at best only partially solve that problem, due to significant overlap in the classified values of the accept and reject pixels. This would require a number of significant additions to the system, first of all more analysis routines for recognizing and handling the partial cases. As there are many more ways to solve a problem partially than to solve it fully, however, more attention would have to be given to other parts of the program, for example the routine that chooses between different alternatives in deciding which ones to actually instantiate, as well as the search level routines for evaluating overall algorithms and for prioritizing the agenda. In addition, the rules for proposing operators as possible *partial* solutions in the operator expert would also have to be revised or added to, since the conditions for deciding that an operator might be a possible *complete* solution can be more exacting than for deciding on it as a possible partial solution. Here more research will be required simply to understand what the partial conditions should be.

Another addition to the capability of the system which I would like to eventually add would include the incorporation of information about the *costs* of operators, in addition to accuracy and consistency, as part of the evaluation functions for judging algorithm states, and for choosing between subproblem solutions. The basis of these costs was described in some detail in Section 2.9. Beyond this, the addition of some means for backtracking on the Sub-Problem Tree, as well as for incorporating semantic contextual knowledge about the problem domain of the image sample, in order to direct or at least limit the search, would help to make the system as a whole more powerful.

Along this same line, one other possibility for improving the problem solving capabilities of the system as a whole, would be to embed it into a more general problem solving architecture, for example the SOAR architecture developed by Laird, Newell, and Rosenbloom [1987]. While I have constructed the problem solving mechanism in REM in a kind of 'home-grown' fashion on an 'as-needed' basis, the possibility of embedding it into a broader, more powerful one, could very well make it easier to add new capabilities to the system more rapidly, and might offer the possibility of adding a learning mechanism to it at a later date. The kind of sub-goaling that the system does, and the different methods that it adopts in different situations, would map very easily into the sub-goaling and problem space concepts used in SOAR.

One other direction to pursue would be to develop a more production oriented system, once the development system began to be successful at solving realistic problems. This would involve taking the emphasis off the display characteristics and the printed messages of the current program, and instead providing the capability of handling larger image sets, and of operating almost entirely covertly (except for debugging purposes), in order to maximize the speed of finding solutions. It might also require some changes in terms of streamlining the program, modifying code to produce fewer garbage collections, and/or providing a larger memory space in which to work.

Also along the line of thinking about a production mode, there are some hardware changes that could additionally be made to improve the speed of the system. While the current system is fairly fast, in that it can solve the limited range of problems that it does often faster than a human (even with all the display updates and terminal output), there is ample room for improvement in this area. First of all, one could establish a closer connection to the Cytocomputer—for example, to have Lisp send image processing commands directly to the controller software in the Cyto microprocessor, instead of sending C4PL command strings to the C4PL parser. Second, one could make the present one-way connection to the Cyto a two-way connection, so that measurements made on images could be passed back to Lisp directly, rather than having them be written to a file and read by Lisp as I am required to do now. There are also a number of Cyto image processing functions, especially the image measurements (histograms and address extraction), which could be speeded up by hardware improvements.

Other ideas on the hardware side would include porting the system to run on a Lisp machine, such as a Symbolics. Additionally, one might consider modifying the code so that it could be ported to other types of image processing machines, for example an array-based machine such as a GAPP, or a hypercube architecture such as an N-Cube or even a Connection Machine. If we really got to the point where we were comfortable with the *method* used by the system to solve more complicated problems, then letting it run on more powerful machines to search farther into the algorithm space would be a reasonable step to take.

In conclusion, I think this project has demonstrated the viability of attempting to bring more formal, automated methods for search and problem solving, into the area of image algorithm development, in particular for set recognition problems. While there is much more to be done, and many difficult problems to solve, we can make progress here if we work at it, and this offers great potential for the future development of the image processing field as a whole.

Appendix A

Partial List of Example Problems Solved by REM

Trivial:

IDENTITY	Y = (X IDENTITY) $Y = X$
COMPLEMT	Y = (X COMPLEMENT) $Y = X^c$

Threshold-Based:

THRESH_RT	Y = (X THRESHOLD-BAND 32 NIL) $Y = (X\|_{32}^{+})$
THRESH_LT	Y = (X THRESHOLD-BAND NIL 30) $Y = (X\|_{31}^{+})^c$
THRESH_IN	Y = (X THRESHOLD-BAND 16 48) $Y = ((X\|_{16}^{+}) \cap (X\|_{49}^{+})^c)$
THRESH_OU	Y = (X THRESHOLD-BAND 41 19) $Y = ((X\|_{41}^{+}) \cup (X\|_{20}^{+})^c)$
IGREY_RT	Y = (X THRESHOLD-BAND 32 NIL) [Ignore Regions] $Y = (X\|_{34}^{+})$

Infmask-Based:

INFX_LT

Y = (X INFMASK-X-BAND NIL 31)
$Y = (X \cap (X \perp^{x} 32)^{c})$

INFX_IN

Y = (X INFMASK-X-BAND 23 43)
$Y = ((X \perp^{x} 23) \cap (X \perp^{x} 44)^{c})$

INFY_RT

Y = (X INFMASK-Y-BAND 39 NIL)
$Y = (X \perp^{y} 39)$

INFY_OU

Y = (X INFMASK-Y-BAND 47 15)
$Y = ((X \perp^{y} 47) \cup (X \cap (X \perp^{y} 16)^{c}))$

IINFY_RT

Y = (X INFMASK-Y-BAND 29 NIL) [Ignore Regions]
$Y = (X \perp^{y} 29)$

CINFY_RT

X1 = (X COMPLEMENT),
Y = (X1 INFMASK-Y-BAND 39 NIL)
$Y = (X^{c} \perp^{y} 39)$

Supmask-Based:

SUPX_LT
Y = (X SUPMASK-X-BAND 32 NIL)
$Y = (X \top_x 32)$

SUPX_IN
Y = (X SUPMASK-X-BAND 44 22)
$Y = ((X \top_x 44) \cap (X \cup (X \top_x 23)^c))$

SUPY_RT
Y = (X SUPMASK-Y-BAND NIL 38)
$Y = (X \cup (X \top_y 39)^c)$

SUPY_OU
Y = (X SUPMASK-Y-BAND 16 46)
$Y = ((X \top_y 16) \cup (X \top_y 47)^c)$

CSUPX_IN
X1 = (X COMPLEMENT),
Y = (X1 SUPMASK-X-BAND 44 22)
$Y = ((X^c \top_x 44) \cap (X^c \cup (X^c \top_x 23)^c))$

Erosion-Based:

EDISK_RT Y = (X ERODE-BAND-PDISK 3 NIL)
$Y = (X \ominus Disk_3)$

EOCT_LT Y = (X ERODE-BAND-OCTAGON NIL 3)
$Y = (X \cap (X \ominus Octagon_4)^c)$

ESQR_IN Y = (X ERODE-BAND-SQUARE 3 4)
$Y = ((X \ominus Square_3) \cap (X \ominus Square_5)^c)$

EDIA_OU Y = (X ERODE-BAND-DIAMOND 5 2)
$Y = ((X \ominus Diamond_5) \cup (X \cap (X \ominus Diamond_3)^c))$

CEDSK_RT X1 = (X COMPLEMENT),
Y = (X1 ERODE-BAND-PDISK 3 NIL)
$Y = (X^c \ominus Disk_3)$

TEDISK_RT Y = (X ERODE-BAND-PDISK 3 NIL) [Mask Errors]
$Y = (X \ominus Disk_3)$

TCEDSK_RT X1 = (X COMPLEMENT),
Y = (X1 ERODE-BAND-PDISK 3 NIL) [Mask Errors]
$Y = (X^c \ominus Disk_3)$

Dilation-Based:

DDISK_RT Y = (X DILATE-BAND-PDISK NIL 2)
$Y = (X \cup (X \circledV Disk_3)^c)$

DOCT_LT Y = (X DILATE-BAND-OCTAGON 4 NIL)
$Y = (X \circledV Octagon_4)$

DSQR_IN Y = (X DILATE-BAND-SQUARE 5 2)
$Y = ((X \circledV Square_5) \cap (X \cup (X \circledV Square_3)^c))$

DDIA_OU Y = (X DILATE-BAND-DIAMOND 2 4)
$Y = ((X \circledV Diamond_2) \cup (X \circledV Diamond_5)^c)$

CDDSK_RT X1 = (X COMPLEMENT),
Y = (X1 DILATE-BAND-PDISK NIL 2)
$Y = (X^c \cup (X^c \circledV Disk_3)^c)$

TCDDSK_LT X1 = (X COMPLEMENT),
Y = (X1 DILATE-BAND-PDISK 3 NIL) [Mask Errors]
$Y = (X^c \circledV Disk_3)$

Opening-Based:

ODISK_LT	Y = (X OPEN-BAND-PDISK NIL 2) $Y = (X \cap (X \circledcirc Disk_3)^c)$
OOCT_RT	Y = (X OPEN-BAND-OCTAGON 4 NIL) $Y = (X \circledcirc Octagon_4)$
OSQR_OU	Y = (X OPEN-BAND-SQUARE 5 2) $Y = ((X \circledcirc Square_5) \cup (X \cap (X \circledcirc Square_3)^c))$
ODIA_IN	Y = (X OPEN-BAND-DIAMOND 2 5) $Y = ((X \circledcirc Diamond_2) \cap (X \circledcirc Diamond_6)^c)$
CODSK_LT	X1 = (X COMPLEMENT), Y = (X1 OPEN-BAND-PDISK NIL 2) $Y = (X^c \cap (X^c \circledcirc Disk_3)^c)$
TCODSK_RT	X1 = (X COMPLEMENT), Y = (X1 OPEN-BAND-PDISK 3 NIL) [Mask Errors] $Y = (X^c \circledcirc Disk_3)$

Closing-Based:

CDISK_LT Y = (X CLOSE-BAND-PDISK 3 NIL)
$Y = (X \odot Disk_3)$

COCT_RT Y = (X CLOSE-BAND-OCTAGON NIL 4)
$Y = (X \cup (X \odot Octagon_5)^c)$

CSQR_OU Y = (X CLOSE-BAND-SQUARE 2 3)
$Y = ((X \odot Square_2) \cup (X \odot Square_4)^c)$

CDIA_IN Y = (X CLOSE-BAND-DIAMOND 5 2)
$Y = ((X \odot Diamond_5) \cap (X \cup (X \odot Diamond_3)^c))$

CCDSK_LT X1 = (X COMPLEMENT),
Y = (X1 CLOSE-BAND-PDISK 3 NIL)
$Y = (X^c \odot Disk_3)$

TCDISK_LT Y = (X CLOSE-BAND-PDISK 3 NIL) [Mask Errors]
$Y = (X \odot Disk_3)$

TCCDSK_LT X1 = (X COMPLEMENT),
Y = (X1 CLOSE-BAND-PDISK 3 NIL) [Mask Errors]
$Y = (X^c \odot Disk_3)$

Particle-Based:

EPDSK_RT Y = (X ERODE-CDILATE-BAND-PDISK 4 NIL)
$Y = ((X \textcircled{\scriptsize$\wedge$} Disk_4) \stackrel{\oplus}{\rightarrow} X)$

EPOCT_LT Y = (X ERODE-CDILATE-BAND-OCTAGON NIL 3)
$Y = (X \cap ((X \textcircled{\scriptsize$\wedge$} Octagon_4) \stackrel{\oplus}{\rightarrow} X)^c)$

EPSQR_IN Y = (X ERODE-CDILATE-BAND-SQUARE 3 4)
$Y = (((X \textcircled{\scriptsize$\wedge$} Square_3) \stackrel{\oplus}{\rightarrow} X) \cap ((X \textcircled{\scriptsize$\wedge$} Square_5) \stackrel{\oplus}{\rightarrow} X)^c)$

EPDIA_OU Y = (X ERODE-CDILATE-BAND-DIAMOND 5 2)
$Y = (((X \textcircled{\scriptsize$\wedge$} Diamond_5) \stackrel{\oplus}{\rightarrow} X) \cup$
$(X \cap ((X \textcircled{\scriptsize$\wedge$} Diamond_3) \stackrel{\oplus}{\rightarrow} X)^c))$

RPSQR_RT Y = (X REDUCE-CDILATE-BAND-SQUARE 5 NIL)
$Y = ((X \triangleright Square_5) \stackrel{\oplus}{\rightarrow} X)$

RPDIA_OU Y = (X REDUCE-CDILATE-BAND-DIAMOND 6 2)
$Y = (((X \triangleright Diamond_6) \stackrel{\oplus}{\rightarrow} X) \cup$
$(X \cap ((X \triangleright Diamond_3) \stackrel{\oplus}{\rightarrow} X)^c))$

CRPSQ_RT X1 = (X COMPLEMENT),
Y = (X1 REDUCE-CDILATE-BAND-SQUARE 5 NIL)
$Y = ((X^c \triangleright Square_5) \stackrel{\oplus}{\rightarrow} X^c)$

Hole-Based:

DHDSK_RT Y = (X DILATE-CERODE-BAND-PDISK NIL 4)

$Y = (X \cup ((X \circledvee Disk_5) \overset{\ominus}{\rightarrow} X)^c)$

DHOCT_LT Y = (X DILATE-CERODE-BAND-OCTAGON 3 NIL)

$Y = ((X \circledvee Octagon_3) \overset{\ominus}{\rightarrow} X)$

DHSQR_IN Y = (X DILATE-CERODE-BAND-SQUARE 5 2)

$Y = (((X \circledvee Square_5) \overset{\ominus}{\rightarrow} X) \cap (X \cup ((X \circledvee Square_3) \overset{\ominus}{\rightarrow} X)^c))$

DHDIA_OU Y = (X DILATE-CERODE-BAND-DIAMOND 3 5)

$Y = (((X \circledvee Diamond_3) \overset{\ominus}{\rightarrow} X) \cup ((X \circledvee Diamond_6) \overset{\ominus}{\rightarrow} X)^c)$

XHSQR_RT Y = (X EXPAND-CERODE-BAND-SQUARE NIL 6)

$Y = (X \cup ((X \lhd Square_7) \overset{\ominus}{\rightarrow} X)^c)$

XHDIA_OU Y = (X EXPAND-CERODE-BAND-DIAMOND 3 5)

$Y = (((X \lhd Diamond_3) \overset{\ominus}{\rightarrow} X) \cup ((X \lhd Diamond_6) \overset{\ominus}{\rightarrow} X)^c)$

CXHSQ_RT X1 = (X COMPLEMENT),

Y = (X1 EXPAND-CERODE-BAND-PDISK NIL 6)

$Y = (X^c \cup ((X^c \lhd Square_7) \overset{\ominus}{\rightarrow} X^c)^c)$

Appendix B

Algebraic Definitions of IC Band Operations

This appendix gives the algebraic definitions for several classes of band operations which are defined within the IC target language described in the text. In all cases except for the threshold operator, there are two categories in each class—one for anti-extensive and one for extensive operations. The anti-extensive ones can be used to categorize pixels in the foreground while the extensive ones can be used to categorize pixels in the background. The operators are all *range-based*, in the sense that they select pixels over a range of values of some (usually spatial) measure. There are four types of ranges: right open-ended, left open-ended, inclusive, and exclusive. The definitions for each type of range are listed below. The symbol Ψ_B stands for the band operation class being defined; the symbol Ψ stands for the underlying base operator from which the band operator is defined.

The binary and grey position mask operators effectively window a certain portion of an image based on absolute position or addresses in the X or Y dimensions. The binary and grey pixel band classes include the definitions for the basic morphology operations of erosion, dilation, opening, and closing, along with their residues. Here erosion and opening follow the anti-extensive definitions while dilation and closing follow the extensive definitions. The binary and grey cases have slightly different definitions due to the different notions of set subtraction which exist for these two.

The threshold band definitions are fairly straightforward—they simply specify different types of threshold ranges on a grey level image. The basic binary particle band operators use conditional dilation and erosion to retain entire particles or holes based usually on their size after an erosion (resp. dilation) operation, or their length as determined by a reduce (resp. expand) operation. This notion is extended to other measures of complete particles and holes in the following two sets of definitions, which make use of additional thinning and thickening operations to perform discriminations on the basis of the number of steps required to obtain a homotopic skeleton, or the number of steps needed to completely prune such a skeleton, respectively. The final set of definitions use successive residues to categorize particles or holes in binary images based on the roughness or shape of their contours.

Binary Position Mask Band

Antiextensive:

$$(X\ \Psi_B\ p_1\ Nil) = (X\ \Psi\ p_1)$$

$$(X\ \Psi_B\ Nil\ p_2) = (X \cap (X\ \Psi\ p_2{+}1)^c)$$

$$(X\ \Psi_B\ p_1\ p_2) = ((X\ \Psi\ p_1) \cap (X\ \Psi\ p_2{+}1)^c)$$

$$(X\ \Psi_B\ p_2\ p_1) = ((X\ \Psi\ p_1) \cup (X \cap (X\ \Psi\ p_2{+}1)^c))$$

Extensive:

$$(X\ \Psi_B\ p_1\ Nil) = (X\ \Psi\ p_1)$$

$$(X\ \Psi_B\ Nil\ p_2) = (X \cup (X\ \Psi\ p_2{+}1)^c)$$

$$(X\ \Psi_B\ p_1\ p_2) = ((X\ \Psi\ p_1) \cup (X\ \Psi\ p_2{+}1)^c)$$

$$(X\ \Psi_B\ p_2\ p_1) = ((X\ \Psi\ p_1) \cap (X \cup (X\ \Psi\ p_2{+}1)^c))$$

Grey Position Mask Band

Antiextensive:

$$(X\ \Psi_B\ p_1\ Nil) = (X\ \Psi\ p_1)$$

$$(X\ \Psi_B\ Nil\ p_2) = (X \uplus (X\ \Psi\ p_2{+}1))$$

$$(X\ \Psi_B\ p_1\ p_2) = ((X\ \Psi\ p_1) \uplus (X\ \Psi\ p_2{+}1))$$

$$(X\ \Psi_B\ p_2\ p_1) = ((X\ \Psi\ p_1) \Cap (X \uplus (X\ \Psi\ p_2{+}1)))$$

Extensive:

$$(X\ \Psi_B\ p_1\ Nil) = (X\ \Psi\ p_1)$$

$$(X\ \Psi_B\ Nil\ p_2) = (X \Cap (X\ \Psi\ p_2{+}1)^c)$$

$$(X\ \Psi_B\ p_1\ p_2) = ((X\ \Psi\ p_1) \Cap (X\ \Psi\ p_2{+}1)^c)$$

$$(X\ \Psi_B\ p_2\ p_1) = ((X\ \Psi\ p_1) \uplus (X \Cap (X\ \Psi\ p_2{+}1)^c))$$

Binary Pixel Band

Antiextensive:

$$(X\ \Psi_B\ p_1\ Nil) = (X\ \Psi\ (S\ p_1))$$

$$(X\ \Psi_B\ Nil\ p_2) = (X \cap (X\ \Psi\ (S\ p_2{+}1))^c)$$

$$(X\ \Psi_B\ p_1\ p_2) = ((X\ \Psi\ (S\ p_1)) \cap (X\ \Psi\ (S\ p_2{+}1))^c)$$

$$(X\ \Psi_B\ p_2\ p_1) = ((X\ \Psi\ (S\ p_1)) \cup (X \cap (X\ \Psi\ (S\ p_2{+}1))^c))$$

Extensive:

$$(X\ \Psi_B\ p_1\ Nil) = (X\ \Psi\ (S\ p_1))$$

$$(X\ \Psi_B\ Nil\ p_2) = (X \cup (X\ \Psi\ (S\ p_2{+}1))^c)$$

$$(X\ \Psi_B\ p_1\ p_2) = ((X\ \Psi\ (S\ p_1)) \cup (X\ \Psi\ (S\ p_2{+}1))^c)$$

$$(X\ \Psi_B\ p_2\ p_1) = ((X\ \Psi\ (S\ p_1)) \cap (X \cup (X\ \Psi\ (S\ p_2{+}1))^c))$$

Grey Pixel Band

Antiextensive:

$$(X\ \Psi_B\ p_1\ Nil) = (X\ \Psi\ (S\ p_1))$$

$$(X\ \Psi_B\ Nil\ p_2) = (X \Cup (X\ \Psi\ (S\ p_2{+}1)))$$

$$(X\ \Psi_B\ p_1\ p_2) = ((X\ \Psi\ (S\ p_1)) \Cup (X\ \Psi\ (S\ p_2{+}1)))$$

$$(X\ \Psi_B\ p_2\ p_1) = ((X\ \Psi\ (S\ p_1)) \Cap (X \Cup (X\ \Psi\ (S\ p_2{+}1))))$$

Extensive:

$$(X\ \Psi_B\ p_1\ Nil) = (X\ \Psi\ (S\ p_1))$$

$$(X\ \Psi_B\ Nil\ p_2) = (X \Cap (X\ \Psi\ (S\ p_2{+}1))^c)$$

$$(X\ \Psi_B\ p_1\ p_2) = ((X\ \Psi\ (S\ p_1)) \Cap (X\ \Psi\ (S\ p_2{+}1))^c)$$

$$(X\ \Psi_B\ p_2\ p_1) = ((X\ \Psi\ (S\ p_1)) \Cup (X \Cap (X\ \Psi\ (S\ p_2{+}1))^c))$$

Threshold Band

$$(X\ \Psi_B\ p_1\ Nil) = (X\ \Psi\ p_1)$$

$$(X\ \Psi_B\ Nil\ p_2) = (X\ \Psi\ p_2{+}1)^c$$

$$(X\ \Psi_B\ p_1\ p_2) = ((X\ \Psi\ p_1) \cap (X\ \Psi\ p_2{+}1)^c)$$

$$(X\ \Psi_B\ p_2\ p_1) = ((X\ \Psi\ p_1) \cup (X\ \Psi\ p_2{+}1)^c)$$

Binary Particle Band (Basic)

Antiextensive:

$$(X\ \Psi_B\ p_1\ Nil) = ((X\ \Psi\ (S\ p_1)) \overset{\oplus}{\rightarrow} X)$$

$$(X\ \Psi_B\ Nil\ p_2) = (X \cap ((X\ \Psi\ (S\ p_2{+}1)) \overset{\oplus}{\rightarrow} X)^c)$$

$$(X\ \Psi_B\ p_1\ p_2) = (((X\ \Psi\ (S\ p_1)) \overset{\oplus}{\rightarrow} X) \cap ((X\ \Psi\ (S\ p_2{+}1)) \overset{\oplus}{\rightarrow} X)^c)$$

$$(X\ \Psi_B\ p_2\ p_1) = (((X\ \Psi\ (S\ p_1)) \overset{\oplus}{\rightarrow} X) \cup (X \cap ((X\ \Psi\ (S\ p_2{+}1)) \overset{\oplus}{\rightarrow} X)^c))$$

Extensive:

$$(X\ \Psi_B\ p_1\ Nil) = ((X\ \Psi\ (S\ p_1)) \overset{\ominus}{\rightarrow} X)$$

$$(X\ \Psi_B\ Nil\ p_2) = (X \cup ((X\ \Psi\ (S\ p_2{+}1)) \overset{\ominus}{\rightarrow} X)^c)$$

$$(X\ \Psi_B\ p_1\ p_2) = (((X\ \Psi\ (S\ p_1)) \overset{\ominus}{\rightarrow} X) \cup ((X\ \Psi\ (S\ p_2{+}1)) \overset{\ominus}{\rightarrow} X)^c)$$

$$(X\ \Psi_B\ p_2\ p_1) = (((X\ \Psi\ (S\ p_1)) \overset{\ominus}{\rightarrow} X) \cap (X \cup ((X\ \Psi\ (S\ p_2{+}1)) \overset{\ominus}{\rightarrow} X)^c))$$

Binary Particle Band (Steps to Digital Skeleton)

Antiextensive:

$$(X\ \Psi_B\ p_1\ Nil) = (((X\ \Psi\ (S\ p_1)) \cap (X \bigcirc (S\ \infty))^c) \overset{\oplus}{\rightarrow} X)$$

$$(X\ \Psi_B\ Nil\ p_2) = (X \cap (((X\ \Psi\ (S\ p_2{+}1)) \cap (X \bigcirc (S\ \infty))^c) \overset{\oplus}{\rightarrow} X)^c)$$

$$(X\ \Psi_B\ p_1\ p_2) = ((((X\ \Psi\ (S\ p_1)) \cap (X \bigcirc (S\ \infty))^c) \overset{\oplus}{\rightarrow} X) \cap (((X\ \Psi\ (S\ p_2{+}1)) \cap (X \bigcirc (S\ \infty))^c) \overset{\oplus}{\rightarrow} X)^c)$$

$$(X\ \Psi_B\ p_2\ p_1) = ((((X\ \Psi\ (S\ p_1)) \cap (X \bigcirc (S\ \infty))^c) \overset{\oplus}{\rightarrow} X) \cup (X \cap (((X\ \Psi\ (S\ p_2{+}1)) \cap (X \bigcirc (S\ \infty))^c) \overset{\oplus}{\rightarrow} X)^c))$$

Extensive:

$$(X\ \Psi_B\ p_1\ Nil) = (((X\ \Psi\ (S\ p_1)) \cup (X \odot (S\ \infty))^c) \overset{\ominus}{\rightarrow} X)$$

$$(X\ \Psi_B\ Nil\ p_2) = (X \cup (((X\ \Psi\ (S\ p_2{+}1)) \cup (X \odot (S\ \infty))^c) \overset{\ominus}{\rightarrow} X)^c)$$

$$(X\ \Psi_B\ p_1\ p_2) = ((((X\ \Psi\ (S\ p_1)) \cup (X \odot (S\ \infty))^c) \overset{\ominus}{\rightarrow} X) \cup (((X\ \Psi\ (S\ p_2{+}1)) \cup (X \odot (S\ \infty))^c) \overset{\ominus}{\rightarrow} X)^c)$$

$$(X\ \Psi_B\ p_2\ p_1) = ((((X\ \Psi\ (S\ p_1)) \cup (X \odot (S\ \infty))^c) \overset{\ominus}{\rightarrow} X) \cap (X \cup (((X\ \Psi\ (S\ p_2{+}1)) \cup (X \odot (S\ \infty))^c) \overset{\ominus}{\rightarrow} X)^c))$$

Note: In the expression '$(S\ \infty)$', S refers to the connectivity of the thinning or thickening operation (4, 6 or 8) and ∞ means to continue the operation until the image stops changing.

Binary Particle Band (Pruning of Digital Skeleton)

Antiextensive:

$$(X\ \Psi_B\ p_1\ Nil) = (((X \bigcirc (S\ \infty))\ \Psi\ (S\ p_1)) \stackrel{\oplus}{\rightarrow} X)$$

$$(X\ \Psi_B\ Nil\ p_2) = (X \cap (((X \bigcirc (S\ \infty))\ \Psi\ (S\ p_2{+}1)) \stackrel{\oplus}{\rightarrow} X)^c)$$

$$\begin{aligned}(X\ \Psi_B\ p_1\ p_2) = ((((&X \bigcirc (S\ \infty))\ \Psi\ (S\ p_1)) \stackrel{\oplus}{\rightarrow} X) \cap \\ &(((X \bigcirc (S\ \infty))\ \Psi\ (S\ p_2{+}1)) \stackrel{\oplus}{\rightarrow} X)^c)\end{aligned}$$

$$\begin{aligned}(X\ \Psi_B\ p_2\ p_1) = ((((&X \bigcirc (S\ \infty))\ \Psi\ (S\ p_1)) \stackrel{\oplus}{\rightarrow} X) \cup \\ &(X \cap (((X \bigcirc (S\ \infty))\ \Psi\ (S\ p_2{+}1)) \stackrel{\oplus}{\rightarrow} X)^c))\end{aligned}$$

Extensive:

$$(X\ \Psi_B\ p_1\ Nil) = (((X \bigodot (S\ \infty))\ \Psi\ (S\ p_1)) \stackrel{\ominus}{\rightarrow} X)$$

$$(X\ \Psi_B\ Nil\ p_2) = (X \cup (((X \bigodot (S\ \infty))\ \Psi\ (S\ p_2{+}1)) \stackrel{\ominus}{\rightarrow} X)^c)$$

$$\begin{aligned}(X\ \Psi_B\ p_1\ p_2) = ((((&X \bigodot (S\ \infty))\ \Psi\ (S\ p_1)) \stackrel{\ominus}{\rightarrow} X) \cup \\ &(((X \bigodot (S\ \infty))\ \Psi\ (S\ p_2{+}1)) \stackrel{\ominus}{\rightarrow} X)^c)\end{aligned}$$

$$\begin{aligned}(X\ \Psi_B\ p_2\ p_1) = ((((&X \bigodot (S\ \infty))\ \Psi\ (S\ p_1)) \stackrel{\ominus}{\rightarrow} X) \cap \\ &(X \cup (((X \bigodot (S\ \infty))\ \Psi\ (S\ p_2{+}1)) \stackrel{\ominus}{\rightarrow} X)^c))\end{aligned}$$

Note: In the expression '$(S\ \infty)$', S refers to the connectivity of the thinning or thickening operation (4, 6 or 8) and ∞ means to continue the operation until the image stops changing.

Binary Particle Band (Residue)

Antiextensive:

$$(X\ \Psi_B\ p_1\ Nil) = (X \cap (((X \cap (X\ \Psi\ (S\ p_1))^c) \ominus \varepsilon) \overset{\oplus}{\rightarrow} X)^c)$$

$$(X\ \Psi_B\ Nil\ p_2) = (((X \cap (X\ \Psi\ (S\ p_2{+}1))^c) \ominus \varepsilon) \overset{\oplus}{\rightarrow} X)$$

$$\begin{aligned}(X\ \Psi_B\ p_1\ p_2) = (&(((X \cap (X\ \Psi\ (S\ p_1))^c) \ominus \varepsilon) \overset{\oplus}{\rightarrow} X)^c \cap \\ &(((X \cap (X\ \Psi\ (S\ p_2{+}1))^c) \ominus \varepsilon) \overset{\oplus}{\rightarrow} X))\end{aligned}$$

$$\begin{aligned}(X\ \Psi_B\ p_2\ p_1) = (&(X \cap (((X \cap (X\ \Psi\ (S\ p_1))^c) \ominus \varepsilon) \overset{\oplus}{\rightarrow} X)^c) \cup \\ &(((X \cap (X\ \Psi\ (S\ p_2{+}1))^c) \ominus \varepsilon) \overset{\oplus}{\rightarrow} X))\end{aligned}$$

Extensive:

$$(X\ \Psi_B\ p_1\ Nil) = (X \cup (((X \cup (X\ \Psi\ (S\ p_1))^c) \oplus \varepsilon) \overset{\ominus}{\rightarrow} X)^c)$$

$$(X\ \Psi_B\ Nil\ p_2) = (((X \cup (X\ \Psi\ (S\ p_2{+}1))^c) \oplus \varepsilon) \overset{\ominus}{\rightarrow} X)$$

$$\begin{aligned}(X\ \Psi_B\ p_1\ p_2) = (&(((X \cup (X\ \Psi\ (S\ p_1))^c) \oplus \varepsilon) \overset{\ominus}{\rightarrow} X)^c \cup \\ &(((X \cup (X\ \Psi\ (S\ p_2{+}1))^c) \oplus \varepsilon) \overset{\ominus}{\rightarrow} X))\end{aligned}$$

$$\begin{aligned}(X\ \Psi_B\ p_2\ p_1) = (&(X \cup (((X \cup (X\ \Psi\ (S\ p_1))^c) \oplus \varepsilon) \overset{\ominus}{\rightarrow} X)^c) \cap \\ &(((X \cup (X\ \Psi\ (S\ p_2{+}1))^c) \oplus \varepsilon) \overset{\ominus}{\rightarrow} X))\end{aligned}$$

Note: The 'ε' symbol in these formulas is a 2x2 square used in these operations to remove insignificant residues which are due primarily to digitization issues.

Appendix C

Terminal Output for Example Problem 'TCODSK_RT'

```
Dribbling to UD$2:[VOGT.REM]DRIBBLE.TEX;1
Lisp>
(R-E-M)
Setting up REM data objects and C4PL image objects
Declaring and Loading Stage Codes
GDECLARE DCYL1 DCYL2 DCYL3 DCYL4 DCYL5 DCYL6 DCYL7 DCYL8
GDECLARE ECYL1 ECYL2 ECYL3 ECYL4 ECYL5 ECYL6 ECYL7 ECYL8
LOADCODE "dcyl1.noc" -> DCYL1
LOADCODE "dcyl2.noc" -> DCYL2
LOADCODE "dcyl3.noc" -> DCYL3
LOADCODE "dcyl4.noc" -> DCYL4
LOADCODE "dcyl5.noc" -> DCYL5
LOADCODE "dcyl6.noc" -> DCYL6
LOADCODE "dcyl7.noc" -> DCYL7
LOADCODE "ecyl1.noc" -> ECYL1
LOADCODE "ecyl2.noc" -> ECYL2
LOADCODE "ecyl3.noc" -> ECYL3
LOADCODE "ecyl4.noc" -> ECYL4
LOADCODE "ecyl5.noc" -> ECYL5
LOADCODE "ecyl6.noc" -> ECYL6
LOADCODE "ecyl7.noc" -> ECYL7
GDECLARE SKTTS1 SKTTS2 SKTTS3 SKTTS4 SKTTS5 SKTTS6 SKTTS7
GDECLARE SKTTD1 SKTTD2 SKTTD3 SKTTD4 SKTTD5 SKTTD6 SKTTD7
LOADCODE "skttsqr1.noc" -> SKTTS1
LOADCODE "skttsqr2.noc" -> SKTTS2
LOADCODE "skttsqr3.noc" -> SKTTS3
LOADCODE "skttsqr4.noc" -> SKTTS4
LOADCODE "skttsqr5.noc" -> SKTTS5
LOADCODE "skttsqr6.noc" -> SKTTS6
LOADCODE "skttsqr7.noc" -> SKTTS7
LOADCODE "skttdia1.noc" -> SKTTD1
LOADCODE "skttdia2.noc" -> SKTTD2
LOADCODE "skttdia3.noc" -> SKTTD3
LOADCODE "skttdia4.noc" -> SKTTD4
LOADCODE "skttdia5.noc" -> SKTTD5
```

```
LOADCODE "skttdia6.noc" -> SKTTD6
LOADCODE "skttdia7.noc" -> SKTTD7
Loading look-up table arrays
GREY
GDECLARE LUTX LUTZ LUT0 LUTE LUTA_S
GDECLARE LUTA_T LUTR_S LUTR_T LUTA_TC
GDECLARE LUTA_T8 LUTR_T8
LOADARRAY "lutx.lut" -> LUTX
LOADARRAY "lutz.lut" -> LUTZ
LOADARRAY "lut0.lut" -> LUT0
LOADARRAY "lute.lut" -> LUTE
LOADARRAY "luta_s.lut" -> LUTA_S
LOADARRAY "luta_t.lut" -> LUTA_T
LOADARRAY "lutr_s.lut" -> LUTR_S
LOADARRAY "lutr_t.lut" -> LUTR_T
LOADARRAY "luta_tc.lut" -> LUTA_TC
LOADARRAY "luta_t8.lut" -> LUTA_T8
LOADARRAY "lutr_t8.lut" -> LUTR_T8

Declaring Image Register and Display Objects
GDECLARE OVERVIEW
GDECLARE MATTE MATTE1 MATTE2 MATTE3 MATTE4
ERASE 256 256 -> MATTE
ERASE 480 480 -> OVERVIEW
ERASE 64 64 -> SCRATCH
GDECLARE MAG4 A_DIST R_DIST A_LINES R_LINES
GDECLARE X0_1 X0_2 X0_3 X0_4
GDECLARE A0_1 A0_2 A0_3 A0_4
GDECLARE R0_1 R0_2 R0_3 R0_4
GDECLARE X_1 X_2 X_3 X_4 A_1 A_2 A_3 A_4 R_1 R_2 R_3 R_4
GDECLARE T_1 T_2 T_3 T_4 T_5 T_6 T_7 T_8 T_9
WINDOW 64 64 29 33 MATTE -> MATTE1
WINDOW 64 64 29 161 MATTE -> MATTE2
WINDOW 64 64 150 33 MATTE -> MATTE3
WINDOW 64 64 150 161 MATTE -> MATTE4
WINDOW 256 256 9 9 OVERVIEW -> MAG4
WINDOW 128 256 1 1 MAG4 -> A_DIST
WINDOW 128 256 129 1 MAG4 -> R_DIST
WINDOW 64 64 272 6 OVERVIEW -> X0_1
WINDOW 64 64 272 72 OVERVIEW -> X0_2
WINDOW 64 64 272 138 OVERVIEW -> X0_3
WINDOW 64 64 272 204 OVERVIEW -> X0_4
WINDOW 64 64 338 6 OVERVIEW -> A0_1
WINDOW 64 64 338 72 OVERVIEW -> A0_2
WINDOW 64 64 338 138 OVERVIEW -> A0_3
```

```
WINDOW 64 64 338 204 OVERVIEW -> A0_4
WINDOW 64 64 404 6 OVERVIEW -> R0_1
WINDOW 64 64 404 72 OVERVIEW -> R0_2
WINDOW 64 64 404 138 OVERVIEW -> R0_3
WINDOW 64 64 404 204 OVERVIEW -> R0_4
WINDOW 64 64 272 272 OVERVIEW -> T_1
WINDOW 64 64 272 338 OVERVIEW -> T_2
WINDOW 64 64 272 404 OVERVIEW -> T_3
WINDOW 64 64 338 272 OVERVIEW -> T_4
WINDOW 64 64 338 338 OVERVIEW -> T_5
WINDOW 64 64 338 404 OVERVIEW -> T_6
WINDOW 64 64 404 272 OVERVIEW -> T_7
WINDOW 64 64 404 338 OVERVIEW -> T_8
WINDOW 64 64 404 404 OVERVIEW -> T_9
WINDOW 64 64 6 272 OVERVIEW -> X_1
WINDOW 64 64 6 338 OVERVIEW -> A_1
WINDOW 64 64 6 404 OVERVIEW -> R_1
WINDOW 64 64 72 272 OVERVIEW -> X_2
WINDOW 64 64 72 338 OVERVIEW -> A_2
WINDOW 64 64 72 404 OVERVIEW -> R_2
WINDOW 64 64 138 272 OVERVIEW -> X_3
WINDOW 64 64 138 338 OVERVIEW -> A_3
WINDOW 64 64 138 404 OVERVIEW -> R_3
WINDOW 64 64 204 272 OVERVIEW -> X_4
WINDOW 64 64 204 338 OVERVIEW -> A_4
WINDOW 64 64 204 404 OVERVIEW -> R_4
DLEVEL := 1

Creating System Display (Overview)
ACTIVATE OVERVIEW
GDECLARE OVLINES
LOADARRAY "ovlines.lut" -> OVLINES
DRAWLINES OVLINES
Creating Image Problem Display (Matte)
ACTIVATE MATTE
COVER 0 216
DRAWLINE 252 1 128 256 128
DRAWLINE 252 121 1 121 256
LOADLUT LUT0 LUT0 LUT0
Initial Setup Complete--Ready to Start


Welcome to REM, a Lisp program that attempts to solve
  simple image processing problems automatically.
The program takes as input a set of example images,
```

```
  along with two disjoint binary (truth) masks that
  indicate the pixel sets to be distinguished (the Accept
  set and the Reject set).  It returns as output an
  algorithm which computes the correct Accept set for each
  example image to within some degree of accuracy (if one
  can be found).
Each image set is composed of four truthed images.


Enter Image Set Name for Problem to Be Solved (or 'exit'):
TCODSK_RT

Enter Tolerance (0 to 1000): 20


IMAGE_SEARCH := "ud$2:[vogt.rem.img]TCODSK_RT#.img"
Initializing Global data structures
ACTIVATE MATTE
DLEVEL := 0
ERASE 64 64 -> MATTE1
ERASE 64 64 -> MATTE2
ERASE 64 64 -> MATTE3
ERASE 64 64 -> MATTE4
COVER 255 216 MATTE -> MATTE
DLEVEL := 1
ACTIVATE OVERVIEW
COVER ~252 0 OVERVIEW -> OVERVIEW
ACTIVATE MATTE
CAPTION 10 10 "TCODSK_RT" 255 -> MATTE


The next few displays illustrate the problem to be solved.
Hit the Return key to advance through them:


LOADLUT LUTX LUTX LUTX
UNSAVE 1 -> MATTE1
UNSAVE 2 -> MATTE2
UNSAVE 3 -> MATTE3
UNSAVE 4 -> MATTE4
Here are the four original grey level images:


LOADLUT LUTR_S LUTA_S LUTO
Here is the desired (binary) result, with the pixels
```

```
  to accept in Green, and those to reject in Red.
  Black areas (if any) are ignored:


LOADLUT LUTR_T LUTA_T LUTZ
Here the Green and Red areas are transparently
  projected onto the originals:



LOADLUT LUTR_S LUTA_S LUTE
Here the Blue channel is used to indicate errors.
  Accept set errors (Misses) tend towards Cyan (from
  Green), while Reject set errors (FAs) tend toward
  Magenta (from Red):


LOADLUT LUTR_T LUTA_T LUTA_TC
In this representation the Reject set is colored from
  Black to Red indicating the degree of error, while
  the Accept set is colored from Green to Blue as a
  a function of the error.

Now hit return and we will try to solve this problem:


Loading Original Images and Masks
DLEVEL := 0
ACTIVATE OVERVIEW
Copying Original Image 1
COPY MATTE1 -> SCRATCH
MAG 4 4 0 1 1 SCRATCH -> MAG4
Splitting Apart Image 1
COPY SCRATCH -> XO_1
BITCLR 7 XO_1 -> XO_1
BITCLR 6 XO_1 -> XO_1
LOADLUT LUTR_S LUTA_S LUTO
ANDIMAGES SCRATCH 128 -> AO_1
COVER 128 63 AO_1 -> AO_1
ANDIMAGES SCRATCH 64 -> RO_1
COVER 64 63 RO_1 -> RO_1
LOADLUT LUTR_T LUTA_T LUTA_TC

Copying Original Image 2
COPY MATTE2 -> SCRATCH
```

```
MAG 4 4 0 1 1 SCRATCH -> MAG4
Splitting Apart Image 2
COPY SCRATCH -> X0_2
BITCLR 7 X0_2 -> X0_2
BITCLR 6 X0_2 -> X0_2
LOADLUT LUTR_S LUTA_S LUT0
ANDIMAGES SCRATCH 128 -> A0_2
COVER 128 63 A0_2 -> A0_2
ANDIMAGES SCRATCH 64 -> R0_2
COVER 64 63 R0_2 -> R0_2
LOADLUT LUTR_T LUTA_T LUTA_TC


Copying Original Image 3
COPY MATTE3 -> SCRATCH
MAG 4 4 0 1 1 SCRATCH -> MAG4
Splitting Apart Image 3
COPY SCRATCH -> X0_3
BITCLR 7 X0_3 -> X0_3
BITCLR 6 X0_3 -> X0_3
LOADLUT LUTR_S LUTA_S LUT0
ANDIMAGES SCRATCH 128 -> A0_3
COVER 128 63 A0_3 -> A0_3
ANDIMAGES SCRATCH 64 -> R0_3
COVER 64 63 R0_3 -> R0_3
LOADLUT LUTR_T LUTA_T LUTA_TC

Copying Original Image 4
COPY MATTE4 -> SCRATCH
MAG 4 4 0 1 1 SCRATCH -> MAG4
Splitting Apart Image 4
COPY SCRATCH -> X0_4
BITCLR 7 X0_4 -> X0_4
BITCLR 6 X0_4 -> X0_4
LOADLUT LUTR_S LUTA_S LUT0
ANDIMAGES SCRATCH 128 -> A0_4
COVER 128 63 A0_4 -> A0_4
ANDIMAGES SCRATCH 64 -> R0_4
COVER 64 63 R0_4 -> R0_4
LOADLUT LUTR_T LUTA_T LUTA_TC

DLEVEL := 1
Finished Loading Original Problem TCODSK_RT
DLEVEL := 1
ACTIVATE MATTE
```

```
ERASE 64 64 -> SCRATCH
Making Initial Problem Evaluation
(Ai Ri) Percentages in X:
NTYPEHIST MATTE1 "hist.dat" (7 881)
POINT 191 64 8 -> SCRATCH
NTYPEHIST MATTE2 "hist.dat" (6 857)
POINT 191 64 10 -> SCRATCH
NTYPEHIST MATTE3 "hist.dat" (5 879)
POINT 191 64 9 -> SCRATCH
NTYPEHIST MATTE4 "hist.dat" (5 832)
POINT 191 64 12 -> SCRATCH
Summarizing Evaluation Results
(Ai Ri) Scores:
(-327 881)  (-312 857)  (-328 879)  (-297 832)
Avg-A = -316  Avg-R = 862
Worst-A = -328  Worst-R = 832
Cent-A = 6  Cent-R = 862
Offset = 866  SD = 20  Consistency = 972


POINT 127 64 10 -> SCRATCH
ACTIVATE OVERVIEW
DLEVEL := 0
MAG 4 4 0 1 1 SCRATCH -> MAG4
CAPTION 10 10 "TCODSK_RT" 255 -> OVERVIEW
DLEVEL := 1

Finished Initial Evaluation
Hit Return to continue:


Prioritizing Agenda
Expanding Next Problem Nodes
Attempting to Solve Problem: PROB0
Calculating X-A-R Images for Problem: PROB0
LOADLUT LUTR_T LUTA_T LUTA_TC
DLEVEL := 0

(Y = X0_1)
($ASSIGN (QUOTE X0_1) (QUOTE T_1))
COPY X0_1 -> T_1
COPY T_1 -> X_1

(Y = A0_1)
($ASSIGN (QUOTE A0_1) (QUOTE T_1))
```

```
COPY A0_1 -> T_1
COPY T_1 -> A_1

(Y = R0_1)
($ASSIGN (QUOTE R0_1) (QUOTE T_1))
COPY R0_1 -> T_1
COPY T_1 -> R_1

(Y = X0_2)
($ASSIGN (QUOTE X0_2) (QUOTE T_1))
COPY X0_2 -> T_1
COPY T_1 -> X_2

(Y = A0_2)
($ASSIGN (QUOTE A0_2) (QUOTE T_1))
COPY A0_2 -> T_1
COPY T_1 -> A_2

(Y = R0_2)
($ASSIGN (QUOTE R0_2) (QUOTE T_1))
COPY R0_2 -> T_1
COPY T_1 -> R_2


(Y = X0_3)
($ASSIGN (QUOTE X0_3) (QUOTE T_1))
COPY X0_3 -> T_1
COPY T_1 -> X_3

(Y = A0_3)
($ASSIGN (QUOTE A0_3) (QUOTE T_1))
COPY A0_3 -> T_1
COPY T_1 -> A_3

(Y = R0_3)
($ASSIGN (QUOTE R0_3) (QUOTE T_1))
COPY R0_3 -> T_1
COPY T_1 -> R_3

(Y = X0_4)
($ASSIGN (QUOTE X0_4) (QUOTE T_1))
COPY X0_4 -> T_1
COPY T_1 -> X_4

(Y = A0_4)
```

```
($ASSIGN (QUOTE AO_4) (QUOTE T_1))
COPY AO_4 -> T_1
COPY T_1 -> A_4

(Y = RO_4)
($ASSIGN (QUOTE RO_4) (QUOTE T_1))
COPY RO_4 -> T_1
COPY T_1 -> R_4
DLEVEL := 1
Finished Calculation of Current Problem X-A-R Images
Hit Return to Continue:


Proposing Operators for Problem PROBO
Test Image Set is : (3 2)
X is Binary

X is Not A-Sup

X is Not R-Inf

X is Not A-Inf

X is Not R-Sup

Return to Continue:




X Does Not Equal A

X Does Not Equal R

Proposed Operators:
REVISE-PROB
Hit Return to Continue:



Starting Data Acquisition for PROBO
Analyzing Proposals for Possible Solutions
Making Choices from List of Candidates
Dispatching Choices and Updating Trees
```

```
Evaluating New Algorithms
Testing for Termination
Prioritizing Agenda
Expanding Next Problem Nodes
Attempting to Solve Problem: PROB1
Calculating X-A-R Images for Problem: PROB1
LOADLUT LUTR_T LUTA_T LUTA_TC

(Y = XO_1)
($ASSIGN (QUOTE XO_1) (QUOTE T_1))
COPY XO_1 -> T_1
COPY T_1 -> X_1

(Y = (AO_1 INTERSECT (XO_1 COMPLEMENT)))
($COMPLEMENT (QUOTE XO_1) (QUOTE T_1))
OUTMSK := 63
NOTIMAGE XO_1 -> T_1
OUTMSK := 255
($INTERSECT (QUOTE AO_1) (QUOTE T_1) (QUOTE T_2))
MINIMAGES AO_1 T_1 -> T_2
COPY T_2 -> A_1

(Y = RO_1)
($ASSIGN (QUOTE RO_1) (QUOTE T_2))
COPY RO_1 -> T_2
COPY T_2 -> R_1

(Y = XO_2)
($ASSIGN (QUOTE XO_2) (QUOTE T_2))
COPY XO_2 -> T_2
COPY T_2 -> X_2


(Y = (AO_2 INTERSECT (XO_2 COMPLEMENT)))
($COMPLEMENT (QUOTE XO_2) (QUOTE T_2))
OUTMSK := 63
NOTIMAGE XO_2 -> T_2
OUTMSK := 255
($INTERSECT (QUOTE AO_2) (QUOTE T_2) (QUOTE T_1))
MINIMAGES AO_2 T_2 -> T_1
COPY T_1 -> A_2

(Y = RO_2)
($ASSIGN (QUOTE RO_2) (QUOTE T_1))
COPY RO_2 -> T_1
```

```
COPY T_1 -> R_2

(Y = X0_3)
($ASSIGN (QUOTE X0_3) (QUOTE T_1))
COPY X0_3 -> T_1
COPY T_1 -> X_3

(Y = (A0_3 INTERSECT (X0_3 COMPLEMENT)))
($COMPLEMENT (QUOTE X0_3) (QUOTE T_1))
OUTMSK := 63
NOTIMAGE X0_3 -> T_1
OUTMSK := 255
($INTERSECT (QUOTE A0_3) (QUOTE T_1) (QUOTE T_2))
MINIMAGES A0_3 T_1 -> T_2
COPY T_2 -> A_3

(Y = R0_3)
($ASSIGN (QUOTE R0_3) (QUOTE T_2))
COPY R0_3 -> T_2
COPY T_2 -> R_3

(Y = X0_4)
($ASSIGN (QUOTE X0_4) (QUOTE T_2))
COPY X0_4 -> T_2
COPY T_2 -> X_4

(Y = (A0_4 INTERSECT (X0_4 COMPLEMENT)))
($COMPLEMENT (QUOTE X0_4) (QUOTE T_2))
OUTMSK := 63
NOTIMAGE X0_4 -> T_2
OUTMSK := 255
($INTERSECT (QUOTE A0_4) (QUOTE T_2) (QUOTE T_1))
MINIMAGES A0_4 T_2 -> T_1
COPY T_1 -> A_4


(Y = R0_4)
($ASSIGN (QUOTE R0_4) (QUOTE T_1))
COPY R0_4 -> T_1
COPY T_1 -> R_4
Finished Calculation of Current Problem X-A-R Images
Hit Return to Continue:


ERASE 64 64 -> SCRATCH
```

```
ACTIVATE MATTE
Evaluating Current Problem Status

COPY X_1 -> MATTE1
DLEVEL := 0
COVER 63 128 A_1 -> A_1
BITMERGE MATTE1 A_1 7 -> MATTE1
COVER 128 63 A_1 -> A_1
COVER 63 64 R_1 -> R_1
BITMERGE MATTE1 R_1 6 -> MATTE1
COVER 64 63 R_1 -> R_1
DLEVEL := 1
COPY X_2 -> MATTE2
DLEVEL := 0
COVER 63 128 A_2 -> A_2
BITMERGE MATTE2 A_2 7 -> MATTE2
COVER 128 63 A_2 -> A_2
COVER 63 64 R_2 -> R_2
BITMERGE MATTE2 R_2 6 -> MATTE2
COVER 64 63 R_2 -> R_2
DLEVEL := 1
COPY X_3 -> MATTE3
DLEVEL := 0
COVER 63 128 A_3 -> A_3
BITMERGE MATTE3 A_3 7 -> MATTE3
COVER 128 63 A_3 -> A_3
COVER 63 64 R_3 -> R_3
BITMERGE MATTE3 R_3 6 -> MATTE3
COVER 64 63 R_3 -> R_3
DLEVEL := 1
COPY X_4 -> MATTE4
DLEVEL := 0
COVER 63 128 A_4 -> A_4
BITMERGE MATTE4 A_4 7 -> MATTE4
COVER 128 63 A_4 -> A_4
COVER 63 64 R_4 -> R_4
BITMERGE MATTE4 R_4 6 -> MATTE4
COVER 64 63 R_4 -> R_4
DLEVEL := 1


New Subproblem to Solve
Hit Return to continue:
```

```
POINT 127 64 10 -> SCRATCH
ACTIVATE OVERVIEW
DLEVEL := 0
MAG 4 4 0 1 1 SCRATCH -> MAG4
CAPTION 10 10 "TCODSK_RT" 255 -> OVERVIEW
DLEVEL := 1

Finished Problem Evaluation
Hit Return to continue:


Proposing Operators for Problem PROB1
Test Image Set is : (3 2)
X is Binary

X is Not A-Sup

X is R-Inf

X is Not A-Inf

X is Not R-Sup

Return to Continue:



X Does Not Equal A

X Does Not Equal R

Proposed Operators:
COMPLEMENT
Hit Return to Continue:



Starting Data Acquisition for PROB1
Analyzing Proposals for Possible Solutions
Making Choices from List of Candidates
Dispatching Choices and Updating Trees


Evaluating New Algorithms
Testing for Termination
```

```
Prioritizing Agenda
Expanding Next Problem Nodes
Attempting to Solve Problem: PROB2
Calculating X-A-R Images for Problem: PROB2
LOADLUT LUTR_T LUTA_T LUTA_TC

(Y = (X0_1 COMPLEMENT))
($COMPLEMENT (QUOTE X0_1) (QUOTE T_1))
OUTMSK := 63
NOTIMAGE X0_1 -> T_1
OUTMSK := 255
COPY T_1 -> X_1

(Y = (A0_1 INTERSECT (X0_1 COMPLEMENT)))
($COMPLEMENT (QUOTE X0_1) (QUOTE T_1))
OUTMSK := 63
NOTIMAGE X0_1 -> T_1
OUTMSK := 255
($INTERSECT (QUOTE A0_1) (QUOTE T_1) (QUOTE T_2))
MINIMAGES A0_1 T_1 -> T_2
COPY T_2 -> A_1

(Y = R0_1)
($ASSIGN (QUOTE R0_1) (QUOTE T_2))
COPY R0_1 -> T_2
COPY T_2 -> R_1

(Y = (X0_2 COMPLEMENT))
($COMPLEMENT (QUOTE X0_2) (QUOTE T_2))
OUTMSK := 63
NOTIMAGE X0_2 -> T_2
OUTMSK := 255
COPY T_2 -> X_2

(Y = (A0_2 INTERSECT (X0_2 COMPLEMENT)))
($COMPLEMENT (QUOTE X0_2) (QUOTE T_2))
OUTMSK := 63
NOTIMAGE X0_2 -> T_2
OUTMSK := 255
($INTERSECT (QUOTE A0_2) (QUOTE T_2) (QUOTE T_1))
MINIMAGES A0_2 T_2 -> T_1
COPY T_1 -> A_2

(Y = R0_2)
($ASSIGN (QUOTE R0_2) (QUOTE T_1))
```

```
COPY RO_2 -> T_1
COPY T_1 -> R_2


(Y = (XO_3 COMPLEMENT))
($COMPLEMENT (QUOTE XO_3) (QUOTE T_1))
OUTMSK := 63
NOTIMAGE XO_3 -> T_1
OUTMSK := 255
COPY T_1 -> X_3

(Y = (AO_3 INTERSECT (XO_3 COMPLEMENT)))
($COMPLEMENT (QUOTE XO_3) (QUOTE T_1))
OUTMSK := 63
NOTIMAGE XO_3 -> T_1
OUTMSK := 255
($INTERSECT (QUOTE AO_3) (QUOTE T_1) (QUOTE T_2))
MINIMAGES AO_3 T_1 -> T_2
COPY T_2 -> A_3

(Y = RO_3)
($ASSIGN (QUOTE RO_3) (QUOTE T_2))
COPY RO_3 -> T_2
COPY T_2 -> R_3

(Y = (XO_4 COMPLEMENT))
($COMPLEMENT (QUOTE XO_4) (QUOTE T_2))
OUTMSK := 63
NOTIMAGE XO_4 -> T_2
OUTMSK := 255
COPY T_2 -> X_4

(Y = (AO_4 INTERSECT (XO_4 COMPLEMENT)))
($COMPLEMENT (QUOTE XO_4) (QUOTE T_2))
OUTMSK := 63
NOTIMAGE XO_4 -> T_2
OUTMSK := 255
($INTERSECT (QUOTE AO_4) (QUOTE T_2) (QUOTE T_1))
MINIMAGES AO_4 T_2 -> T_1
COPY T_1 -> A_4

(Y = RO_4)
($ASSIGN (QUOTE RO_4) (QUOTE T_1))
COPY RO_4 -> T_1
COPY T_1 -> R_4
```

```
Finished Calculation of Current Problem X-A-R Images
Hit Return to Continue:


ROTATE 180 SCRATCH -> SCRATCH
ACTIVATE MATTE
Evaluating Current Problem Status


COPY X_1 -> MATTE1
DLEVEL := 0
COVER 63 128 A_1 -> A_1
BITMERGE MATTE1 A_1 7 -> MATTE1
COVER 128 63 A_1 -> A_1
COVER 63 64 R_1 -> R_1
BITMERGE MATTE1 R_1 6 -> MATTE1
COVER 64 63 R_1 -> R_1
DLEVEL := 1
COPY X_2 -> MATTE2
DLEVEL := 0
COVER 63 128 A_2 -> A_2
BITMERGE MATTE2 A_2 7 -> MATTE2
COVER 128 63 A_2 -> A_2
COVER 63 64 R_2 -> R_2
BITMERGE MATTE2 R_2 6 -> MATTE2
COVER 64 63 R_2 -> R_2
DLEVEL := 1
COPY X_3 -> MATTE3
DLEVEL := 0
COVER 63 128 A_3 -> A_3
BITMERGE MATTE3 A_3 7 -> MATTE3
COVER 128 63 A_3 -> A_3
COVER 63 64 R_3 -> R_3
BITMERGE MATTE3 R_3 6 -> MATTE3
COVER 64 63 R_3 -> R_3
DLEVEL := 1
COPY X_4 -> MATTE4
DLEVEL := 0
COVER 63 128 A_4 -> A_4
BITMERGE MATTE4 A_4 7 -> MATTE4
COVER 128 63 A_4 -> A_4
COVER 63 64 R_4 -> R_4
BITMERGE MATTE4 R_4 6 -> MATTE4
COVER 64 63 R_4 -> R_4
DLEVEL := 1
```

```
New Subproblem to Solve
Hit Return to continue:


POINT 127 1 55 -> SCRATCH
ACTIVATE OVERVIEW
DLEVEL := 0
MAG 4 4 0 1 1 SCRATCH -> MAG4
CAPTION 10 10 "TCODSK_RT" 255 -> OVERVIEW
DLEVEL := 1


Finished Problem Evaluation
Hit Return to continue:


Proposing Operators for Problem PROB2
Test Image Set is : (4 3)
X is Binary

X is A-Sup

X is Not R-Inf

X is Not A-Inf

X is Not R-Sup

Return to Continue:



X Does Not Equal A

X Does Not Equal R

Testing for X is A-Sup-Xaddr
($INTERSECT (QUOTE R_4) (QUOTE X_4) (QUOTE T_2))
MINIMAGES R_4 X_4 -> T_2
($DILATE (QUOTE A_4) (QUOTE LINE-NS) 63 (QUOTE T_3))
MAX3D 20 20 20 63 # A_4 -> T_3
($DILATE (QUOTE T_2) (QUOTE LINE-NS) 63 (QUOTE T_4))
MAX3D 20 20 20 63 # T_2 -> T_4
($INTERSECT (QUOTE T_3) (QUOTE T_4) (QUOTE T_5))
```

```
MINIMAGES T_3 T_4 -> T_5
($ERODE (QUOTE T_5) (QUOTE LINE-EW) 2 (QUOTE T_4))
MIN3D 0 222 0 2 # T_5 -> T_4
COPY T_4 -> T_1
NZPCOUNT T_1 "count.dat"

X is Not A-Sup-Xaddr


Testing for X is A-Sup-Yaddr
($INTERSECT (QUOTE R_4) (QUOTE X_4) (QUOTE T_2))
MINIMAGES R_4 X_4 -> T_2
($DILATE (QUOTE A_4) (QUOTE LINE-EW) 63 (QUOTE T_4))
MAX3D 0 222 0 63 # A_4 -> T_4
($DILATE (QUOTE T_2) (QUOTE LINE-EW) 63 (QUOTE T_5))
MAX3D 0 222 0 63 # T_2 -> T_5
($INTERSECT (QUOTE T_4) (QUOTE T_5) (QUOTE T_3))
MINIMAGES T_4 T_5 -> T_3
($ERODE (QUOTE T_3) (QUOTE LINE-NS) 2 (QUOTE T_5))
MIN3D 20 20 20 2 # T_3 -> T_5
COPY T_5 -> T_1
NZPCOUNT T_1 "count.dat"

X is Not A-Sup-Yaddr


Testing for X is A-Sup-1NoMixParts
Testing for X is A-Sup-AllNoMixParts
Testing for X is A-Sup-AllMixBords
Testing for X is A-Sup-NoMixBords
($COMPLEMENT (QUOTE A_4) (QUOTE T_6))
OUTMSK := 63
NOTIMAGE A_4 -> T_6
OUTMSK := 255
($INTERSECT (QUOTE X_4) (QUOTE T_6) (QUOTE T_7))
MINIMAGES X_4 T_6 -> T_7
COPY T_7 -> T_1
($CDILATE (QUOTE A_4) (QUOTE X_4) (QUOTE T_7))
MINIMAGES A_4 X_4 -> T_7
COVER 63 31 T_7 -> T_7
COVER 63 32 X_4 -> X_4
ADDIMAGES T_7 X_4 -> T_7
COVER 32 63 X_4 -> X_4
SPAN 63 32 63 64 # # # T_7 -> T_7
COVER 32 0 T_7 -> T_7
COPY T_7 -> T_2
```

```
($CDILATE (QUOTE T_1) (QUOTE X_4) (QUOTE T_7))
MINIMAGES T_1 X_4 -> T_7
COVER 63 31 T_7 -> T_7
COVER 63 32 X_4 -> X_4
ADDIMAGES T_7 X_4 -> T_7
COVER 32 63 X_4 -> X_4
SPAN 63 32 63 64 # # # T_7 -> T_7
COVER 32 0 T_7 -> T_7
COPY T_7 -> T_5
($INTERSECT (QUOTE T_2) (QUOTE T_5) (QUOTE T_7))
MINIMAGES T_2 T_5 -> T_7
COPY T_7 -> T_3
NZPCOUNT T_3 "count.dat"
($COMPLEMENT (QUOTE T_3) (QUOTE T_7))
OUTMSK := 63
NOTIMAGE T_3 -> T_7
OUTMSK := 255
($INTERSECT (QUOTE T_2) (QUOTE T_7) (QUOTE T_6))
MINIMAGES T_2 T_7 -> T_6
COPY T_6 -> T_2
($COMPLEMENT (QUOTE T_3) (QUOTE T_6))
OUTMSK := 63
NOTIMAGE T_3 -> T_6
OUTMSK := 255
($INTERSECT (QUOTE T_5) (QUOTE T_6) (QUOTE T_7))
MINIMAGES T_5 T_6 -> T_7
COPY T_7 -> T_5
NZPCOUNT T_2 "count.dat"
NZPCOUNT T_5 "count.dat"
($ERODE (QUOTE X_4) (QUOTE DIAMOND) 1 (QUOTE T_7))
MIN3D 20 222 20 1 # X_4 -> T_7
($COMPLEMENT (QUOTE T_7) (QUOTE T_6))
OUTMSK := 63
NOTIMAGE T_7 -> T_6
OUTMSK := 255
($INTERSECT (QUOTE X_4) (QUOTE T_6) (QUOTE T_7))
MINIMAGES X_4 T_6 -> T_7
($INTERSECT (QUOTE T_3) (QUOTE T_7) (QUOTE T_6))
MINIMAGES T_3 T_7 -> T_6
COPY T_6 -> T_4
($INTERSECT (QUOTE A_4) (QUOTE T_3) (QUOTE T_6))
MINIMAGES A_4 T_3 -> T_6
($COMPLEMENT (QUOTE T_6) (QUOTE T_7))
OUTMSK := 63
NOTIMAGE T_6 -> T_7
```

```
OUTMSK := 255
($INTERSECT (QUOTE T_4) (QUOTE T_7) (QUOTE T_6))
MINIMAGES T_4 T_7 -> T_6
COPY T_6 -> T_2
($INTERSECT (QUOTE T_1) (QUOTE T_3) (QUOTE T_6))
MINIMAGES T_1 T_3 -> T_6
($COMPLEMENT (QUOTE T_6) (QUOTE T_7))
OUTMSK := 63
NOTIMAGE T_6 -> T_7
OUTMSK := 255
($INTERSECT (QUOTE T_4) (QUOTE T_7) (QUOTE T_6))
MINIMAGES T_4 T_7 -> T_6
COPY T_6 -> T_5
NZPCOUNT T_2 "count.dat"
NZPCOUNT T_5 "count.dat"
NZPCOUNT T_2 "count.dat"
NZPCOUNT T_5 "count.dat"

($COMPLEMENT (QUOTE A_3) (QUOTE T_6))
OUTMSK := 63
NOTIMAGE A_3 -> T_6
OUTMSK := 255
($INTERSECT (QUOTE X_3) (QUOTE T_6) (QUOTE T_7))
MINIMAGES X_3 T_6 -> T_7
COPY T_7 -> T_1
($CDILATE (QUOTE A_3) (QUOTE X_3) (QUOTE T_7))
MINIMAGES A_3 X_3 -> T_7
COVER 63 31 T_7 -> T_7
COVER 63 32 X_3 -> X_3
ADDIMAGES T_7 X_3 -> T_7
COVER 32 63 X_3 -> X_3
SPAN 63 32 63 64 # # # T_7 -> T_7
COVER 32 0 T_7 -> T_7
COPY T_7 -> T_2
($CDILATE (QUOTE T_1) (QUOTE X_3) (QUOTE T_7))
MINIMAGES T_1 X_3 -> T_7
COVER 63 31 T_7 -> T_7
COVER 63 32 X_3 -> X_3
ADDIMAGES T_7 X_3 -> T_7
COVER 32 63 X_3 -> X_3
SPAN 63 32 63 64 # # # T_7 -> T_7
COVER 32 0 T_7 -> T_7
COPY T_7 -> T_5
($INTERSECT (QUOTE T_2) (QUOTE T_5) (QUOTE T_7))
MINIMAGES T_2 T_5 -> T_7
```

```
COPY T_7 -> T_3
($ERODE (QUOTE X_3) (QUOTE DIAMOND) 1 (QUOTE T_7))
MIN3D 20 222 20 1 # X_3 -> T_7
($COMPLEMENT (QUOTE T_7) (QUOTE T_6))
OUTMSK := 63
NOTIMAGE T_7 -> T_6
OUTMSK := 255
($INTERSECT (QUOTE X_3) (QUOTE T_6) (QUOTE T_7))
MINIMAGES X_3 T_6 -> T_7
($INTERSECT (QUOTE T_3) (QUOTE T_7) (QUOTE T_6))
MINIMAGES T_3 T_7 -> T_6
COPY T_6 -> T_4
($INTERSECT (QUOTE A_3) (QUOTE T_3) (QUOTE T_6))
MINIMAGES A_3 T_3 -> T_6
($COMPLEMENT (QUOTE T_6) (QUOTE T_7))
OUTMSK := 63
NOTIMAGE T_6 -> T_7
OUTMSK := 255
($INTERSECT (QUOTE T_4) (QUOTE T_7) (QUOTE T_6))
MINIMAGES T_4 T_7 -> T_6
COPY T_6 -> T_2
($INTERSECT (QUOTE T_1) (QUOTE T_3) (QUOTE T_6))
MINIMAGES T_1 T_3 -> T_6
($COMPLEMENT (QUOTE T_6) (QUOTE T_7))
OUTMSK := 63
NOTIMAGE T_6 -> T_7
OUTMSK := 255
($INTERSECT (QUOTE T_4) (QUOTE T_7) (QUOTE T_6))
MINIMAGES T_4 T_7 -> T_6
COPY T_6 -> T_5
NZPCOUNT T_2 "count.dat"
NZPCOUNT T_5 "count.dat"


X is A-Sup-1NoMixParts

X is Not A-Sup-AllNoMixParts

X is A-Sup-AllMixBords

X is Not A-Sup-NoMixBords

Return to Continue:
```

```
Proposed Operators:
OPEN-BAND-PDISK
Hit Return to Continue:



Starting Data Acquisition for PROB2
Collecting Data for Operation OPEN-BAND-PDISK
Coding into Register T_1
RAINBOW 1 8
RAINBOW 55 62
DLEVEL := 0
COVER ~0 0 T_1 -> T_1
COPY X_1 -> T_2
COVER ~0 1 T_2 -> T_2
ADDIMAGES T_2 T_1 FALSE -> T_1
APPLY ECYL1 1 X_1 -> T_2
APPLY DCYL1 1 T_2 -> T_2
SLICE 1 1 63 0 T_1 -> T_5
DIFFIMAGES T_2 T_5 FALSE -> T_5
OUTMSK := 63
NOTIMAGE T_5 -> T_5
OUTMSK := 255
COVER ~0 1 T_2 -> T_2
ADDIMAGES T_2 T_1 FALSE -> T_1
MINIMAGES T_1 T_5 -> T_1
APPLY ECYL2 1 X_1 -> T_2
APPLY DCYL2 1 T_2 -> T_2
SLICE 2 2 63 0 T_1 -> T_5
DIFFIMAGES T_2 T_5 FALSE -> T_5
OUTMSK := 63
NOTIMAGE T_5 -> T_5
OUTMSK := 255
COVER ~0 1 T_2 -> T_2
ADDIMAGES T_2 T_1 FALSE -> T_1
MINIMAGES T_1 T_5 -> T_1
APPLY ECYL3 1 X_1 -> T_2
APPLY DCYL3 1 T_2 -> T_2
SLICE 3 3 63 0 T_1 -> T_5
DIFFIMAGES T_2 T_5 FALSE -> T_5
OUTMSK := 63
NOTIMAGE T_5 -> T_5
OUTMSK := 255
COVER ~0 1 T_2 -> T_2
```

```
ADDIMAGES T_2 T_1 FALSE -> T_1
MINIMAGES T_1 T_5 -> T_1
APPLY ECYL4 1 X_1 -> T_2
APPLY DCYL4 1 T_2 -> T_2
SLICE 4 4 63 0 T_1 -> T_5
DIFFIMAGES T_2 T_5 FALSE -> T_5
OUTMSK := 63
NOTIMAGE T_5 -> T_5
OUTMSK := 255
COVER ~0 1 T_2 -> T_2
ADDIMAGES T_2 T_1 FALSE -> T_1
MINIMAGES T_1 T_5 -> T_1
APPLY ECYL5 1 X_1 -> T_2
APPLY DCYL5 1 T_2 -> T_2
SLICE 5 5 63 0 T_1 -> T_5
DIFFIMAGES T_2 T_5 FALSE -> T_5
OUTMSK := 63
NOTIMAGE T_5 -> T_5
OUTMSK := 255
COVER ~0 1 T_2 -> T_2
ADDIMAGES T_2 T_1 FALSE -> T_1
MINIMAGES T_1 T_5 -> T_1
APPLY ECYL6 1 X_1 -> T_2
APPLY DCYL6 1 T_2 -> T_2
SLICE 6 6 63 0 T_1 -> T_5
DIFFIMAGES T_2 T_5 FALSE -> T_5
OUTMSK := 63
NOTIMAGE T_5 -> T_5
OUTMSK := 255
COVER ~0 1 T_2 -> T_2
ADDIMAGES T_2 T_1 FALSE -> T_1
MINIMAGES T_1 T_5 -> T_1
APPLY ECYL7 1 X_1 -> T_2
APPLY DCYL7 1 T_2 -> T_2
SLICE 7 7 63 0 T_1 -> T_5
DIFFIMAGES T_2 T_5 FALSE -> T_5
OUTMSK := 63
NOTIMAGE T_5 -> T_5
OUTMSK := 255
COVER ~0 1 T_2 -> T_2
ADDIMAGES T_2 T_1 FALSE -> T_1
MINIMAGES T_1 T_5 -> T_1
MAG 4 4 0 1 1 T_1 -> MAG4
CAPTION 10 10 "TCODSK_RT" 255 -> OVERVIEW
DLEVEL := 1
```

```
Coded Image Number 1
Return to Continue:



DLEVEL := 0
LOADLUT LUTR_T8 LUTA_T8 LUTZ
COVER 63 128 A_1 -> A_1
BITMERGE T_1 A_1 7 -> T_1
COVER 128 63 A_1 -> A_1
COVER 63 64 R_1 -> R_1
BITMERGE T_1 R_1 6 -> T_1
COVER 64 63 R_1 -> R_1
MAG 4 4 0 1 1 T_1 -> MAG4
CAPTION 10 10 "TCODSK_RT" 255 -> OVERVIEW
DLEVEL := 1
Overlaid Coded Image 1
Return to Continue:


Computing A-R Dists for Instance 1
NTYPEHIST T_1 "hist.dat"
Drawing A-R Dists
DLEVEL := 0
ERASE 256 256 -> MAG4
LOADARRAY "a_lines.dat" -> A_LINES
LOADARRAY "r_lines.dat" -> R_LINES
DRAWLINES A_LINES 4 A_DIST -> A_DIST
DRAWLINES R_LINES 4 R_DIST -> R_DIST
CAPTION 10 10 "TCODSK_RT" 255 -> OVERVIEW
DLEVEL := 1
A-R Dists for Coded Image 1
Return to Continue:



RAINBOW 1 8
RAINBOW 55 62
DLEVEL := 0
COVER ~0 0 T_1 -> T_1
COPY X_2 -> T_2
COVER ~0 1 T_2 -> T_2
ADDIMAGES T_2 T_1 FALSE -> T_1
APPLY ECYL1 1 X_2 -> T_2
APPLY DCYL1 1 T_2 -> T_2
SLICE 1 1 63 0 T_1 -> T_5
```

```
DIFFIMAGES T_2 T_5 FALSE -> T_5
OUTMSK := 63
NOTIMAGE T_5 -> T_5
OUTMSK := 255
COVER ~0 1 T_2 -> T_2
ADDIMAGES T_2 T_1 FALSE -> T_1
MINIMAGES T_1 T_5 -> T_1
APPLY ECYL2 1 X_2 -> T_2
APPLY DCYL2 1 T_2 -> T_2
SLICE 2 2 63 0 T_1 -> T_5
DIFFIMAGES T_2 T_5 FALSE -> T_5
OUTMSK := 63
NOTIMAGE T_5 -> T_5
OUTMSK := 255
COVER ~0 1 T_2 -> T_2
ADDIMAGES T_2 T_1 FALSE -> T_1
MINIMAGES T_1 T_5 -> T_1
APPLY ECYL3 1 X_2 -> T_2
APPLY DCYL3 1 T_2 -> T_2
SLICE 3 3 63 0 T_1 -> T_5
DIFFIMAGES T_2 T_5 FALSE -> T_5
OUTMSK := 63
NOTIMAGE T_5 -> T_5
OUTMSK := 255
COVER ~0 1 T_2 -> T_2
ADDIMAGES T_2 T_1 FALSE -> T_1
MINIMAGES T_1 T_5 -> T_1
APPLY ECYL4 1 X_2 -> T_2
APPLY DCYL4 1 T_2 -> T_2
SLICE 4 4 63 0 T_1 -> T_5
DIFFIMAGES T_2 T_5 FALSE -> T_5
OUTMSK := 63
NOTIMAGE T_5 -> T_5
OUTMSK := 255
COVER ~0 1 T_2 -> T_2
ADDIMAGES T_2 T_1 FALSE -> T_1
MINIMAGES T_1 T_5 -> T_1
APPLY ECYL5 1 X_2 -> T_2
APPLY DCYL5 1 T_2 -> T_2
SLICE 5 5 63 0 T_1 -> T_5
DIFFIMAGES T_2 T_5 FALSE -> T_5
OUTMSK := 63
NOTIMAGE T_5 -> T_5
OUTMSK := 255
COVER ~0 1 T_2 -> T_2
```

```
ADDIMAGES T_2 T_1 FALSE -> T_1
MINIMAGES T_1 T_5 -> T_1
APPLY ECYL6 1 X_2 -> T_2
APPLY DCYL6 1 T_2 -> T_2
SLICE 6 6 63 0 T_1 -> T_5
DIFFIMAGES T_2 T_5 FALSE -> T_5
OUTMSK := 63
NOTIMAGE T_5 -> T_5
OUTMSK := 255
COVER ~0 1 T_2 -> T_2
ADDIMAGES T_2 T_1 FALSE -> T_1
MINIMAGES T_1 T_5 -> T_1
APPLY ECYL7 1 X_2 -> T_2
APPLY DCYL7 1 T_2 -> T_2
SLICE 7 7 63 0 T_1 -> T_5
DIFFIMAGES T_2 T_5 FALSE -> T_5
OUTMSK := 63
NOTIMAGE T_5 -> T_5
OUTMSK := 255
COVER ~0 1 T_2 -> T_2
ADDIMAGES T_2 T_1 FALSE -> T_1
MINIMAGES T_1 T_5 -> T_1
MAG 4 4 0 1 1 T_1 -> MAG4
CAPTION 10 10 "TCODSK_RT" 255 -> OVERVIEW
DLEVEL := 1
Coded Image Number 2
DLEVEL := 0
LOADLUT LUTR_T8 LUTA_T8 LUTZ
COVER 63 128 A_2 -> A_2
BITMERGE T_1 A_2 7 -> T_1
COVER 128 63 A_2 -> A_2
COVER 63 64 R_2 -> R_2
BITMERGE T_1 R_2 6 -> T_1
COVER 64 63 R_2 -> R_2
MAG 4 4 0 1 1 T_1 -> MAG4
CAPTION 10 10 "TCODSK_RT" 255 -> OVERVIEW
DLEVEL := 1
Overlaid Coded Image 2
Computing A-R Dists for Instance 2
NTYPEHIST T_1 "hist.dat"
Drawing A-R Dists
DLEVEL := 0
ERASE 256 256 -> MAG4
LOADARRAY "a_lines.dat" -> A_LINES
LOADARRAY "r_lines.dat" -> R_LINES
```

```
DRAWLINES A_LINES 4 A_DIST -> A_DIST
DRAWLINES R_LINES 4 R_DIST -> R_DIST
CAPTION 10 10 "TCODSK_RT" 255 -> OVERVIEW
DLEVEL := 1
A-R Dists for Coded Image 2

RAINBOW 1 8
RAINBOW 55 62
DLEVEL := 0
COVER ~0 0 T_1 -> T_1
COPY X_3 -> T_2
COVER ~0 1 T_2 -> T_2
ADDIMAGES T_2 T_1 FALSE -> T_1
APPLY ECYL1 1 X_3 -> T_2
APPLY DCYL1 1 T_2 -> T_2
SLICE 1 1 63 0 T_1 -> T_5
DIFFIMAGES T_2 T_5 FALSE -> T_5
OUTMSK := 63
NOTIMAGE T_5 -> T_5
OUTMSK := 255
COVER ~0 1 T_2 -> T_2
ADDIMAGES T_2 T_1 FALSE -> T_1
MINIMAGES T_1 T_5 -> T_1
APPLY ECYL2 1 X_3 -> T_2
APPLY DCYL2 1 T_2 -> T_2
SLICE 2 2 63 0 T_1 -> T_5
DIFFIMAGES T_2 T_5 FALSE -> T_5
OUTMSK := 63
NOTIMAGE T_5 -> T_5
OUTMSK := 255
COVER ~0 1 T_2 -> T_2
ADDIMAGES T_2 T_1 FALSE -> T_1
MINIMAGES T_1 T_5 -> T_1
APPLY ECYL3 1 X_3 -> T_2
APPLY DCYL3 1 T_2 -> T_2
SLICE 3 3 63 0 T_1 -> T_5
DIFFIMAGES T_2 T_5 FALSE -> T_5
OUTMSK := 63
NOTIMAGE T_5 -> T_5
OUTMSK := 255
COVER ~0 1 T_2 -> T_2
ADDIMAGES T_2 T_1 FALSE -> T_1
MINIMAGES T_1 T_5 -> T_1
APPLY ECYL4 1 X_3 -> T_2
APPLY DCYL4 1 T_2 -> T_2
```

```
SLICE 4 4 63 0 T_1 -> T_5
DIFFIMAGES T_2 T_5 FALSE -> T_5
OUTMSK := 63
NOTIMAGE T_5 -> T_5
OUTMSK := 255
COVER ~0 1 T_2 -> T_2
ADDIMAGES T_2 T_1 FALSE -> T_1
MINIMAGES T_1 T_5 -> T_1
APPLY ECYL5 1 X_3 -> T_2
APPLY DCYL5 1 T_2 -> T_2
SLICE 5 5 63 0 T_1 -> T_5
DIFFIMAGES T_2 T_5 FALSE -> T_5
OUTMSK := 63
NOTIMAGE T_5 -> T_5
OUTMSK := 255
COVER ~0 1 T_2 -> T_2
ADDIMAGES T_2 T_1 FALSE -> T_1
MINIMAGES T_1 T_5 -> T_1
APPLY ECYL6 1 X_3 -> T_2
APPLY DCYL6 1 T_2 -> T_2
SLICE 6 6 63 0 T_1 -> T_5
DIFFIMAGES T_2 T_5 FALSE -> T_5
OUTMSK := 63
NOTIMAGE T_5 -> T_5
OUTMSK := 255
COVER ~0 1 T_2 -> T_2
ADDIMAGES T_2 T_1 FALSE -> T_1
MINIMAGES T_1 T_5 -> T_1
APPLY ECYL7 1 X_3 -> T_2
APPLY DCYL7 1 T_2 -> T_2
SLICE 7 7 63 0 T_1 -> T_5
DIFFIMAGES T_2 T_5 FALSE -> T_5
OUTMSK := 63
NOTIMAGE T_5 -> T_5
OUTMSK := 255
COVER ~0 1 T_2 -> T_2
ADDIMAGES T_2 T_1 FALSE -> T_1
MINIMAGES T_1 T_5 -> T_1
MAG 4 4 0 1 1 T_1 -> MAG4
CAPTION 10 10 "TCODSK_RT" 255 -> OVERVIEW
DLEVEL := 1
Coded Image Number 3
DLEVEL := 0
LOADLUT LUTR_T8 LUTA_T8 LUTZ
COVER 63 128 A_3 -> A_3
```

```
BITMERGE T_1 A_3 7 -> T_1
COVER 128 63 A_3 -> A_3
COVER 63 64 R_3 -> R_3
BITMERGE T_1 R_3 6 -> T_1
COVER 64 63 R_3 -> R_3
MAG 4 4 0 1 1 T_1 -> MAG4
CAPTION 10 10 "TCODSK_RT" 255 -> OVERVIEW
DLEVEL := 1
Overlaid Coded Image 3
Computing A-R Dists for Instance 3
NTYPEHIST T_1 "hist.dat"
Drawing A-R Dists
DLEVEL := 0
ERASE 256 256 -> MAG4
LOADARRAY "a_lines.dat" -> A_LINES
LOADARRAY "r_lines.dat" -> R_LINES
DRAWLINES A_LINES 4 A_DIST -> A_DIST
DRAWLINES R_LINES 4 R_DIST -> R_DIST
CAPTION 10 10 "TCODSK_RT" 255 -> OVERVIEW
DLEVEL := 1
A-R Dists for Coded Image 3

RAINBOW 1 8
RAINBOW 55 62
DLEVEL := 0
COVER ~0 0 T_1 -> T_1
COPY X_4 -> T_2
COVER ~0 1 T_2 -> T_2
ADDIMAGES T_2 T_1 FALSE -> T_1
APPLY ECYL1 1 X_4 -> T_2
APPLY DCYL1 1 T_2 -> T_2
SLICE 1 1 63 0 T_1 -> T_5
DIFFIMAGES T_2 T_5 FALSE -> T_5
OUTMSK := 63
NOTIMAGE T_5 -> T_5
OUTMSK := 255
COVER ~0 1 T_2 -> T_2
ADDIMAGES T_2 T_1 FALSE -> T_1
MINIMAGES T_1 T_5 -> T_1
APPLY ECYL2 1 X_4 -> T_2
APPLY DCYL2 1 T_2 -> T_2
SLICE 2 2 63 0 T_1 -> T_5
DIFFIMAGES T_2 T_5 FALSE -> T_5
OUTMSK := 63
NOTIMAGE T_5 -> T_5
```

```
OUTMSK := 255
COVER ~0 1 T_2 -> T_2
ADDIMAGES T_2 T_1 FALSE -> T_1
MINIMAGES T_1 T_5 -> T_1
APPLY ECYL3 1 X_4 -> T_2
APPLY DCYL3 1 T_2 -> T_2
SLICE 3 3 63 0 T_1 -> T_5
DIFFIMAGES T_2 T_5 FALSE -> T_5
OUTMSK := 63
NOTIMAGE T_5 -> T_5
OUTMSK := 255
COVER ~0 1 T_2 -> T_2
ADDIMAGES T_2 T_1 FALSE -> T_1
MINIMAGES T_1 T_5 -> T_1
APPLY ECYL4 1 X_4 -> T_2
APPLY DCYL4 1 T_2 -> T_2
SLICE 4 4 63 0 T_1 -> T_5
DIFFIMAGES T_2 T_5 FALSE -> T_5
OUTMSK := 63
NOTIMAGE T_5 -> T_5
OUTMSK := 255
COVER ~0 1 T_2 -> T_2
ADDIMAGES T_2 T_1 FALSE -> T_1
MINIMAGES T_1 T_5 -> T_1
APPLY ECYL5 1 X_4 -> T_2
APPLY DCYL5 1 T_2 -> T_2
SLICE 5 5 63 0 T_1 -> T_5
DIFFIMAGES T_2 T_5 FALSE -> T_5
OUTMSK := 63
NOTIMAGE T_5 -> T_5
OUTMSK := 255
COVER ~0 1 T_2 -> T_2
ADDIMAGES T_2 T_1 FALSE -> T_1
MINIMAGES T_1 T_5 -> T_1
APPLY ECYL6 1 X_4 -> T_2
APPLY DCYL6 1 T_2 -> T_2
SLICE 6 6 63 0 T_1 -> T_5
DIFFIMAGES T_2 T_5 FALSE -> T_5
OUTMSK := 63
NOTIMAGE T_5 -> T_5
OUTMSK := 255
COVER ~0 1 T_2 -> T_2
ADDIMAGES T_2 T_1 FALSE -> T_1
MINIMAGES T_1 T_5 -> T_1
APPLY ECYL7 1 X_4 -> T_2
```

```
APPLY DCYL7 1 T_2 -> T_2
SLICE 7 7 63 0 T_1 -> T_5
DIFFIMAGES T_2 T_5 FALSE -> T_5
OUTMSK := 63
NOTIMAGE T_5 -> T_5
OUTMSK := 255
COVER ~0 1 T_2 -> T_2
ADDIMAGES T_2 T_1 FALSE -> T_1
MINIMAGES T_1 T_5 -> T_1
MAG 4 4 0 1 1 T_1 -> MAG4
CAPTION 10 10 "TCODSK_RT" 255 -> OVERVIEW
DLEVEL := 1
Coded Image Number 4
DLEVEL := 0
LOADLUT LUTR_T8 LUTA_T8 LUTZ
COVER 63 128 A_4 -> A_4
BITMERGE T_1 A_4 7 -> T_1
COVER 128 63 A_4 -> A_4
COVER 63 64 R_4 -> R_4
BITMERGE T_1 R_4 6 -> T_1
COVER 64 63 R_4 -> R_4
MAG 4 4 0 1 1 T_1 -> MAG4
CAPTION 10 10 "TCODSK_RT" 255 -> OVERVIEW
DLEVEL := 1
Overlaid Coded Image 4
Computing A-R Dists for Instance 4
NTYPEHIST T_1 "hist.dat"
Drawing A-R Dists
DLEVEL := 0
ERASE 256 256 -> MAG4
LOADARRAY "a_lines.dat" -> A_LINES
LOADARRAY "r_lines.dat" -> R_LINES
DRAWLINES A_LINES 4 A_DIST -> A_DIST
DRAWLINES R_LINES 4 R_DIST -> R_DIST
CAPTION 10 10 "TCODSK_RT" 255 -> OVERVIEW
DLEVEL := 1
A-R Dists for Coded Image 4

Analyzing Proposals for Possible Solutions
Calculating Union of Vgrams for OPEN-BAND-PDISK
#*000111110000000000000000000000000000000000000000
#*111000000000000000000000000000000000000000000000
Analyzing Complete Solutions for OPEN-BAND-PDISK
Complete Solution Possible
Evaluating Step Accuracy and Consistency
```

```
(Ai Ri) Percentages in X:
(1000 0)  (1000 0)  (1000 0)  (1000 0)
Summarizing Evaluation Results
(Ai Ri) Scores:
(1000 -414)  (1000 -414)  (1000 -414)  (1000 -414)
Avg-A = 1000  Avg-R = -414
Worst-A = 1000  Worst-R = -414
Cent-A = 1000  Cent-R = 0
Offset = 1000  SD = 0  Consistency = 1000

Making Choices from List of Candidates
Dispatching Choices and Updating Trees

Evaluating New Algorithms
Testing for Termination
Terminating Search for This Problem
LOADLUT LUTR_T LUTA_T LUTA_TC

Solution(s) Found!
Hit Return to List (Next) Solution:



(X1 = (X0 COMPLEMENT))
(Y = (X1 OPEN-BAND-PDISK 3 NIL))

(X1 = (X0 COMPLEMENT))
(Y = (X1 OPEN (PDISK 3)))

Hit Return to Execute and Display Solution:



DLEVEL := 1
ACTIVATE OVERVIEW
($COMPLEMENT (QUOTE X0_1) (QUOTE T_2))
OUTMSK := 63
NOTIMAGE X0_1 -> T_2
OUTMSK := 255
($OPEN (QUOTE T_2) (QUOTE PDISK) 3 (QUOTE T_5))
APPLY ECYL3 1 T_2 -> T_5
APPLY DCYL3 1 T_5 -> T_5
COPY T_5 -> T_1
ACTIVATE MATTE
COPY T_1 -> MATTE1
DLEVEL := 0
```

```
COVER 63 128 A0_1 -> A0_1
BITMERGE MATTE1 A0_1 7 -> MATTE1
COVER 128 63 A0_1 -> A0_1
COVER 63 64 R0_1 -> R0_1
BITMERGE MATTE1 R0_1 6 -> MATTE1
COVER 64 63 R0_1 -> R0_1
DLEVEL := 1

ACTIVATE OVERVIEW
($COMPLEMENT (QUOTE X0_2) (QUOTE T_2))
OUTMSK := 63
NOTIMAGE X0_2 -> T_2
OUTMSK := 255
($OPEN (QUOTE T_2) (QUOTE PDISK) 3 (QUOTE T_5))
APPLY ECYL3 1 T_2 -> T_5
APPLY DCYL3 1 T_5 -> T_5
COPY T_5 -> T_1
ACTIVATE MATTE
COPY T_1 -> MATTE2
DLEVEL := 0
COVER 63 128 A0_2 -> A0_2
BITMERGE MATTE2 A0_2 7 -> MATTE2
COVER 128 63 A0_2 -> A0_2
COVER 63 64 R0_2 -> R0_2
BITMERGE MATTE2 R0_2 6 -> MATTE2
COVER 64 63 R0_2 -> R0_2
DLEVEL := 1


ACTIVATE OVERVIEW
($COMPLEMENT (QUOTE X0_3) (QUOTE T_2))
OUTMSK := 63
NOTIMAGE X0_3 -> T_2
OUTMSK := 255
($OPEN (QUOTE T_2) (QUOTE PDISK) 3 (QUOTE T_5))
APPLY ECYL3 1 T_2 -> T_5
APPLY DCYL3 1 T_5 -> T_5
COPY T_5 -> T_1
ACTIVATE MATTE
COPY T_1 -> MATTE3
DLEVEL := 0
COVER 63 128 A0_3 -> A0_3
BITMERGE MATTE3 A0_3 7 -> MATTE3
COVER 128 63 A0_3 -> A0_3
COVER 63 64 R0_3 -> R0_3
```

```
BITMERGE MATTE3 RO_3 6 -> MATTE3
COVER 64 63 RO_3 -> RO_3
DLEVEL := 1

ACTIVATE OVERVIEW
($COMPLEMENT (QUOTE XO_4) (QUOTE T_2))
OUTMSK := 63
NOTIMAGE XO_4 -> T_2
OUTMSK := 255
($OPEN (QUOTE T_2) (QUOTE PDISK) 3 (QUOTE T_5))
APPLY ECYL3 1 T_2 -> T_5
APPLY DCYL3 1 T_5 -> T_5
COPY T_5 -> T_1
ACTIVATE MATTE
COPY T_1 -> MATTE4
DLEVEL := 0
COVER 63 128 AO_4 -> AO_4
BITMERGE MATTE4 AO_4 7 -> MATTE4
COVER 128 63 AO_4 -> AO_4
COVER 63 64 RO_4 -> RO_4
BITMERGE MATTE4 RO_4 6 -> MATTE4
COVER 64 63 RO_4 -> RO_4
DLEVEL := 1

Results of algorithm on original images
Hit Return to continue:



ACTIVATE OVERVIEW
UNSAVE 5 -> MATTE1
COPY MATTE1 -> XO_1
BITCLR 7 XO_1 -> XO_1
BITCLR 6 XO_1 -> XO_1
ANDIMAGES MATTE1 128 -> AO_1
COVER 128 63 AO_1 -> AO_1
ANDIMAGES MATTE1 64 -> RO_1
COVER 64 63 RO_1 -> RO_1
ACTIVATE OVERVIEW
($COMPLEMENT (QUOTE XO_1) (QUOTE T_2))
OUTMSK := 63
NOTIMAGE XO_1 -> T_2
OUTMSK := 255
($OPEN (QUOTE T_2) (QUOTE PDISK) 3 (QUOTE T_5))
APPLY ECYL3 1 T_2 -> T_5
```

```
APPLY DCYL3 1 T_5 -> T_5
COPY T_5 -> T_1
ACTIVATE MATTE
COPY T_1 -> MATTE1
DLEVEL := 0
COVER 63 128 AO_1 -> AO_1
BITMERGE MATTE1 AO_1 7 -> MATTE1
COVER 128 63 AO_1 -> AO_1
COVER 63 64 RO_1 -> RO_1
BITMERGE MATTE1 RO_1 6 -> MATTE1
COVER 64 63 RO_1 -> RO_1
DLEVEL := 1

Fifth image result at upper left


"Thank You for Your Patience!"


Break in function TERMINATE-SEARCH (signaled with BREAK).
Last Chance to look before we move on.
Break> (CONTINUE)
Continuing from break loop...

Enter Image Set Name for Problem to Be Solved (or 'exit'):
EXIT

NIL
Lisp> (EXIT)
```

Appendix D

Band Operators Defined in the IC Target Language

Set and Threshold Operators:

IDENTITY
COMPLEMENT

UNION
INTERSECT

THRESHOLD-BAND

Binary Position Masks:

INFMASK-X-BAND
INFMASK-Y-BAND
SUPMASK-X-BAND
SUPMASK-Y-BAND

Grey-Level Operators:

GREY-INFMASK-X-BAND
GREY-INFMASK-Y-BAND
GREY-SUPMASK-X-BAND
GREY-SUPMASK-Y-BAND

GREY-ERODE-BAND-PDISK
GREY-OPEN-BAND-PDISK
GREY-DILATE-BAND-PDISK
GREY-CLOSE-BAND-PDISK

Erosion Operators:

ERODE-BAND-PDISK
ERODE-BAND-OCTAGON
ERODE-BAND-SQUARE
ERODE-BAND-DIAMOND
ERODE-BAND-LINE-EW
ERODE-BAND-LINE-NS
ERODE-BAND-LINE-NESW
ERODE-BAND-LINE-NWSE
ERODE-BAND-VECT-N
ERODE-BAND-VECT-NE
ERODE-BAND-VECT-E
ERODE-BAND-VECT-SE
ERODE-BAND-VECT-S
ERODE-BAND-VECT-SW
ERODE-BAND-VECT-W
ERODE-BAND-VECT-NW

Dilation Operators:

DILATE-BAND-PDISK
DILATE-BAND-OCTAGON
DILATE-BAND-SQUARE
DILATE-BAND-DIAMOND
DILATE-BAND-LINE-EW
DILATE-BAND-LINE-NS
DILATE-BAND-LINE-NESW
DILATE-BAND-LINE-NWSE
DILATE-BAND-VECT-N
DILATE-BAND-VECT-NE
DILATE-BAND-VECT-E
DILATE-BAND-VECT-SE
DILATE-BAND-VECT-S
DILATE-BAND-VECT-SW
DILATE-BAND-VECT-W
DILATE-BAND-VECT-NW

Opening Operators:

OPEN-BAND-PDISK
OPEN-BAND-OCTAGON
OPEN-BAND-SQUARE
OPEN-BAND-DIAMOND
OPEN-BAND-LINE-EW
OPEN-BAND-LINE-NS
OPEN-BAND-LINE-NESW
OPEN-BAND-LINE-NWSE

Closing Operators:

CLOSE-BAND-PDISK
CLOSE-BAND-OCTAGON
CLOSE-BAND-SQUARE
CLOSE-BAND-DIAMOND
CLOSE-BAND-LINE-EW
CLOSE-BAND-LINE-NS
CLOSE-BAND-LINE-NESW
CLOSE-BAND-LINE-NWSE

Particle Size and Length Operators:

ERODE-CDILATE-BAND-PDISK
ERODE-CDILATE-BAND-OCTAGON
ERODE-CDILATE-BAND-SQUARE
ERODE-CDILATE-BAND-DIAMOND
ERODE-CDILATE-BAND-LINE-EW
ERODE-CDILATE-BAND-LINE-NS
ERODE-CDILATE-BAND-LINE-NESW
ERODE-CDILATE-BAND-LINE-NWSE

REDUCE-CDILATE-BAND-SQUARE
REDUCE-CDILATE-BAND-DIAMOND

HTHIN-CDILATE-BAND-SQUARE
HTHIN-CDILATE-BAND-DIAMOND

PRUNE-CDILATE-BAND-SQUARE
PRUNE-CDILATE-BAND-DIAMOND

Hole Size and Length Operators:

DILATE-CERODE-BAND-PDISK
DILATE-CERODE-BAND-OCTAGON
DILATE-CERODE-BAND-SQUARE
DILATE-CERODE-BAND-DIAMOND
DILATE-CERODE-BAND-LINE-EW
DILATE-CERODE-BAND-LINE-NS
DILATE-CERODE-BAND-LINE-NESW
DILATE-CERODE-BAND-LINE-NWSE

EXPAND-CERODE-BAND-SQUARE
EXPAND-CERODE-BAND-DIAMOND

HTHICK-CERODE-BAND-SQUARE
HTHICK-CERODE-BAND-DIAMOND

PLUG-CERODE-BAND-SQUARE
PLUG-CERODE-BAND-DIAMOND

Particle Roughness/Shape Operators:

OPENRES-CDILATE-BAND-PDISK
OPENRES-CDILATE-BAND-OCTAGON
OPENRES-CDILATE-BAND-SQUARE
OPENRES-CDILATE-BAND-DIAMOND
OPENRES-CDILATE-BAND-LINE-EW
OPENRES-CDILATE-BAND-LINE-NS
OPENRES-CDILATE-BAND-LINE-NESW
OPENRES-CDILATE-BAND-LINE-NWSE

Hole Roughness/Shape Operators:

CLOSERES-CERODE-BAND-PDISK
CLOSERES-CERODE-BAND-OCTAGON
CLOSERES-CERODE-BAND-SQUARE
CLOSERES-CERODE-BAND-DIAMOND
CLOSERES-CERODE-BAND-LINE-EW
CLOSERES-CERODE-BAND-LINE-NS
CLOSERES-CERODE-BAND-LINE-NESW
CLOSERES-CERODE-BAND-LINE-NWSE

Appendix E

Selected Bibliography

Abbott, L., Haralick, R. M., and Zhuang, X.
"Pipeline Architectures for Morphologic Image Analysis". *Machine Vision and Applications*, v. 1, no. 1, pps. 23–40, 1988.

Balzer, R., Goldman, N., and Wile, D. "Informality in program specification". *IJCAI 5*, pps. 389–397, 1977.

Barstow, D. "A knowledge based system for automatic program construction". *IJCAI 5*, pps. 382–388, 1977.

Barstow, D. *Knowledge-Based Program Construction.* Elsevier press, Amsterdam, 1979.

Crimmins, T. R., and Brown, W. M. "Image Algebra and Automatic Shape Recognition". *IEEE Transactions on Aerospace and Electronic Systems*, v. AES-21, no. 1, January 1985.

Doughtery, E. R., and Giardina, C. R.
Image Processing—Continuous to Discrete, Volume I. Prentice-Hall, Englewood Cliffs, New Jersey, 1987.

Environmental Research Institute of Michigan "C4PL User's Manual, Version 1.1". June 1987.

Giardina, C. R., and Doughtery, E. R. *Morphological Methods in Image and Signal Processing*, Prentice-Hall, Englewood Cliffs, New Jersey, 1988.

Gillies, A. M. *Machine Learning Procedures for Generating Image Domain Feature Detectors.* Doctoral Dissertation, Computer and Communication Sciences Dept., University of Michigan, 1985.

Goetcharian, V. "From Binary to Gray Level Tone Image Processing by Using Fuzzy Logic Concepts". *Pattern Recognition*, v. 12, pps. 7–15, 1980.

Green, C. "A summary of the PSI program synthesis system". *IJCAI 5*, pps. 380–381, 1977.

Green, C., and Barstow, D. "On program synthesis knowledge". *Artificial Intelligence*, v. 3, pps. 241–279, 1978.

Haas, L. "Automatic programming of machine vision systems". *IJCAI 10*, pps. 790–792, 1987.

Haralick, R. M., Sternberg, S. R., and Zhuang, X. "Image Analysis Using Mathematical Morphology". *IEEE Trans. on Pattern Analysis and Machine Intelligence*, v. 9, no. 4, pps. 532–550, July 1987.

Heidorn, G. "Automatic programming through natural language dialog: A survey". *IBM J. Research and Development*, v. 4, pps. 302–313, 1976.

Hoare, C. A. R. "Mathematics of Programming". *BYTE*, pps. 115–126, August 1986.

Holland, J. H. *Adaptation in Natural and Artifical Systems.* University of Michigan Press, Ann Arbor, Michigan, 1975.

Kant, E. *Efficiency Considerations in Program Synthesis: A Knowledge-Based Approach.* Doctoral Dissertation, Computer Science Dept., Stanford University, 1979.

Laird, J. E., Newell, A., and Rosenbloom, P. S. "SOAR: An Architecture for General Intelligence". *Artificial Intelligence*, v. 33, pps. 1–64, 1987.

Langley, P., Simon, H., Bradshaw, G., and Zytkow, J. *Scientific Discovery: Computational Explorations of the Creative Processes.* MIT Press, Cambridge, Massachusetts, 1987.

Lougheed, R. M. "A high-speed recirculating neighborhood architecture". *Proc. SPIE Vol. 534 Architectures and Algorithms for Digital Image Processing II*, Los Angeles, California, pps. 22–33, 1985.

Lougheed, R. M. "Advanced image-processing architectures for machine vision". *Proc. SPIE Vol. 755 Image Pattern Recognition, Los Angeles, California*, pps. 35–51, 1987.

Lougheed, R. M., and Overmars, J. "Description of a high-level environment for image processing algorithm development". *Proc. Intl. Electronic Imaging Exposition and Conf.*, Boston, Massachusetts, 1987.

Lougheed, R. M., and Sampson, R. E. "3-D Imaging Systems and High-Speed Processing for Robot Control". *Machine Vision and Applications*, v. 1, no. 1, pps. 41–58, 1988.

Manna, Z., and Waldinger, R. "DEDALUS–The DEDuctive ALgorithm Ur-Synthesizer". *Proc. National Computer Conf.*, Anaheim, California, pps. 683–690, 1978.

Maragos, P., and Schafer, R. W. "Morphological Skeleton Representation and Coding of Binary Images". *IEEE Trans. on Acoustics, Speech, and Signal Processing*, v. ASSP-34, no. 5, pps. 1228–1244, October 1986.

Maragos, P., and Schafer, R. W. "Morphological Filters—Part I: Their Set-Theoretic Analysis and Relations to Linear Shift-Invariant Filters". *IEEE Trans. on Acoustics, Speech, and Signal Processing*, v. ASSP-35, no. 8, pps. 1153–1169, August 1987a.

Maragos, P., and Schafer, R. W. "Morphological Filters—Part II: Their Relations to Median, Order-Statistic, and Stack Filters". *IEEE Trans. on Acoustics, Speech, and Signal Processing*, v. ASSP-35, no. 8, pps. 1170–1184, August 1987b.

Marr, D. *Vision.* W. H. Freeman, San Francisco, 1982.

Rich, C., and Waters, R. C. "Automatic Programming: Myths and Prospects". *Computer*, v. 21, no. 8, pps. 40–52, August 1988.

Ritter, G. X., and Wilson, J. N. "Image Algebra: A Unified Approach to Image Processing". *SPIE Proc. on Medical Imaging*, Newport Beach, California, 1987.

Ritter, G. X., and Wilson, J. N. "The Image Algebra in A Nutshell". *Proc. First Int. Conf. on Computer Vision*, London, England, 1987.

Sakaue, K., and Tamura, H. "Automatic Generation of Image Processing Programs by Knowledge-based Verification". *IEEE CH2145*, pps. 189–192, 1985.

Scott, P. D., and Vogt, R. C. "Knowledge-Oriented Learning". *IJCAI 11*, Karlsrue, West Germany, August 1983.

Serra, J. *Image Analysis and Mathematical Morphology.* Academic Press, London, 1982.

Serra, J. "Introduction to Mathematical Morphology". *Computer Vision, Graphics, and Image Processing*, v. 35, pps. 283–305, 1986.

Serra, J. (Ed.) *Image Analysis and Mathematical Morphology, Volume 2: Theoretical Advances.* Academic Press, London, 1988.

Shih, F. Y., and Mitchell, O. R. "Decomposition of Gray Scale Morphological Structuring Elements". *IEEE Workshop on Computer Vision*, Miami Beach, Florida, Nov. 1987.

Steele, G. L. *Common LISP: The Language.* Digital Equipment Corporation, 1984.

Sternberg, S. R. "Biomedical Image Processing". *Computer*, v. 16, no. 1, pps. 22–34, January 1983.

Sternberg, S. R. "Grayscale Morphology". *Computer Vision, Graphics, and Image Processing*, v. 35, pps. 333–355, September 1986.

Tomita, F. "Interactive and Automatic Image Recognition System". *Machine Vision and Applications*, v. 1, no. 1, pps. 59–70, 1988.

Vogt, R. C. "Algebraic Languages and Structures for Image Analysis Using Cellular or Neighborhood Operations". Perkin-Elmer Corp., Internal Report, December 1982.

Vogt, R. C. "Formalized Approaches to Image Algorithm Development Using Mathematical Morphology". *Proc. VISION '86*, Detroit, Michigan, June 1986.

Vogt, R. C. "Morphological operator distributions based on monotonicity and the problem posed by digital disk-shaped structuring elements". *Proc. SPIE 938 Digital and Optical Shape Recognition and Pattern Recognition*, Orlando, Florida, April 1988.

Vogt, R. C. "Automatic Generation of Simple Morphological Algorithms". *IEEE Conf. Computer Vision and Pattern Recognition*, Ann Arbor, Michigan, June 1988.

Vogt, R. C. "Precise Extraction of Bone in CT Images of the Head". To appear in *Algorithms in Mathematical Morphology*, Academic Press, London, 1989.

Index